How Fast Did *T. rex* Run?

*Unsolved Questions from the Frontiers of
Dinosaur Science*

DAVID HONE

Princeton University Press
Princeton and Oxford

Published in the United States, Canada, and the Philippines by
Princeton University Press
41 William Street, Princeton, New Jersey 08540
press.princeton.edu

First published in Great Britain in 2022 as *The Future of Dinosaurs* by Hodder & Stoughton
A Hachette UK company

ISBN 978-0-691-24251-4
ISBN (e-book) 978-0-691-24252-1
Library of Congress Control Number 2022930988

Typeset in Bembo by Hewer Text UK Ltd, Edinburgh

Jacket art: T. rex skeleton photograph by Eirik Newth; Triceratops skeleton photograph
by Eva Kröcher; Velociraptor skeleton photograph by Eduard Solà Vázquez

Cover design by Karl Spurzem

Printed in the United States of America

1 3 5 7 9 10 8 6 4 2

For Christine
Not forgotten

Contents

Acknowledgements   ix

Preface   xi

Introduction   1

1. Extinction   11
2. Origins and Relationships   23
3. Preservation   39
4. Diversity   55
5. Evolutionary Patterns   73
6. Habitats and Environments   87
7. Anatomy   97
8. Mechanics and Movement   109
9. Physiology   123
10. Coverings   135
11. Appearance   149
12. Reproduction   163
13. Behaviour   175
14. Ecology   189
15. Dinosaur Descendants   203
16. Research and Communication   211
17. Coda   223

References   239

Index   241

Acknowledgements

Special thanks to Jordan Mallon for reading the whole damned thing and making suggestions and comments on the content, at least one of which I actually followed and changed.

Also thanks to Sarah Labelle, Marissa Livius, Bryan Moore, Mathew Roloson and for discussion and feedback on the book.

Numerous colleagues helped me to obtain photos or provided them for this book and so I also thank Caleb Brown, Sara Burch, Andrea Cau, Julia Clarke, Mick Ellison, David Evans, Peter Falkingham, Pascal Godefroit, Scott Hartman, Donald Henderson, Thierry Hubin, ReBecca Hunt-Foster, Evan Johnson-Ransom, Martin Kundrát, Hans Larsson, Xing Lida, Jordan Mallon, Maria McNamara, Alejandro Otero, Diego Pol, Eric Snively, Larry Witmer, Xu Xing and Matt Zeher.

I also need to thank my agent Max Edwards, and Huw Armstrong, Maddie Price and Barry Johnston at Hodder for shepherding this work from an original idea into its final form.

Preface

CONSIDER THE *TYRANNOSAURUS rex*. This most iconic of dinosaurs is the superstar of movies, features in endless documentaries and appears in every popular book of dinosaurs (generally on the cover) that one could imagine. In your mind's eye, I'm sure there's already a clear picture of this incredible animal and if you know a bit about it, you probably have some details down about it, too. You may perhaps have an idea of its length, weight, height, number of teeth or top speed; if you are particularly keen, you may know of its bite power, skin texture, habitat, favoured prey species, and more.

Much of this, you may imagine, is based on some rigorous science and fossil data, or at least as rigorous as it can be, given that the last *Tyrannosaurus* died in the great mass extinction of the dinosaurs some 65 or so million years ago. Hundreds of papers in the scientific literature have described the sizes and shapes of bones, reconstructed the cartilage of the joints, worked out which muscles would attach where on the skeleton, identified patches of fossil skin, looked at footprints and bite marks on bones, calculated mass estimates and walking speeds, and more. We can put together a remarkably detailed picture of this animal.

More than that, in fact, we can delve into some incredible features of which even fans of dinosaurs would likely be unaware. We can look inside the skulls of *Tyrannosaurus* specimens to see how large the various parts of their brains were, and we can get an estimate of their range of hearing from the structure of the inner ear. There have been studies looking at pupil shape of the eyes, nocturnal versus diurnal habits, growth rate, the sex of individual animals, and how far on average they would have to travel to find food when scavenging.

Putting all of this together gives us an unparalleled picture of an animal that has been extinct for a million human lifetimes. We know more about *Tyrannosaurus* than perhaps any other extinct dinosaur,* but we have only around twenty-five good skeletons to work from, and this creature is just one of some 1,500 or so dinosaur species currently known to science.

Even for the things we do know, we can't compare many of these facts to other species that the *Tyrannosaurus* lived alongside, or that came before it. Yes, it's amazing to have good estimates of the speed of this huge carnivore, but it is also frustrating to be unable to answer questions such as 'could it catch *Triceratops*?' when we don't know how fast they were. Furthermore, our knowledge of rexy is still full of huge holes – we don't know what colour it was or what its eggs or nests looked like. We don't know if it lived in groups, if it mated for life, if it preferred forests or open environments, or if it migrated in winter. For all our technological advances, and two centuries of new data and ideas, we still know less than the basic ecologies of living beings: what parasites and diseases afflicted them, how they communicated, if they ever took fish as prey, what their internal organs were like, or even what their tiny arms were used for.

Whenever I do some kind of outreach or public engagement with science, I encounter things about dinosaurs that the public are absolutely amazed to learn scientists know with certainty, and also things they assume would be easy to work out, for which we have no real idea of the answer. It's a curious quirk that there is such a disparity between what palaeontologists do know and what many people think we know. This book is, therefore, ultimately about what we *don't* know about dinosaurs.

There are major gaps in our knowledge, but extraordinary advances in palaeontological methods and ever more dinosaur fossils promise a landslide of new data and huge leaps forward in our understanding of these incredible creatures. There are a great many issues that we are currently unable to resolve, but we have tantalising hints that we may soon be able to answer them. This book aims to bridge the gap between

---

* The birds are living dinosaurs, something we will deal with later, but for now accept that for the rest of this book the term 'dinosaur' excludes the birds, unless otherwise stated. For the record, the correct technical term for what most people would call a dinosaur is a 'non-avian dinosaur'.

what we do know and what gaps there are in our understanding, but also to examine just how we are likely to fill them in the future.

Ongoing research trends, as yet undescribed specimens and still-developing techniques mean we can plot a route to the next generation of knowledge of dinosaur biology. We *will* make mistakes in the future, and have doubtless made some in the past that have yet to be corrected, but the inexorable progress of scientific discovery will doubtless improve on what we have now.

We know enough to spot some key gaps and have the fossils to try and fill them, but there will be exciting discoveries and some most unexpected results on the way. Based on the last two centuries of dinosaurian research, that much is certain. There are also gaps that may never be filled, or, perhaps worse, we may be able to tentatively fill them, but not know if our calculated answers are correct. We have probably learned more about dinosaurs in the last twenty years than in the previous two hundred, and are poised to take many more steps forwards in the coming decade. This book will address the recent strides made and the advanced knowledge we have of these astonishing creatures, as well as what we hope to learn in the future about these most fascinating of extinct animals.

I'd also add that while this book has been kept as up to date as possible, the field of dinosaur palaeontology is constantly advancing and remains full of contradictions. Evidence is often tantalisingly incomplete and it can mean several ideas are near equally valid, or the weight of data hangs in the balance. I've tried to steer as even a course through this as possible and stick to the more mainstream hypotheses (while recognising some of the more important alternatives or contradictions), but it's impossible to cover every aspect and there will be researchers who disagree most strongly in places with what I have said. Even allowing for that, there will be other issues thanks to the advances made. While writing this book, I've constantly had to update multiple chapters and sections, and doubtless between the time this is signed off by my editor and you read these words, a paper or two will have come out that fills a gap I claim is unfilled, or overturns a hypothesis I had advocated as being correct. This is inevitable, science advances after all, but be warned that for all my best efforts, this book will retain or even develop controversy as research continues.

A common accusation levelled at scientists is that they are always changing their minds, as if this is somehow a negative. The oft-rephrased (and never quite certainly attributed) quote of John Maynard Keynes is most apt: when the facts change, I change my mind. Regardless of the original source or correct wording of this statement, it is of course the correct approach to take. If old analyses are shown to be flawed or problematic, or if new data comes to light (an obviously very common phenomenon in palaeontology), then the weight of evidence can shift.

Given the huge gaps in our knowledge of dinosaurs, it should be little surprise that the weight of evidence will often shift (and on occasion lurch back again) on key subjects, even some where we thought we were confident about the results. It can be frustrating, but it is a sure sign that we are learning and the science is getting better, not worse.

Introduction

IN ANY KIND of description of the history of a scientific field, there will be a fundamental narrative of uncertainly giving way to fact and theory, with unknowns and gaps in our knowledge being filled in and worked out. But perhaps inevitably, for every fact to hand or inference that can be made, there is another that was unknown or uncertain. Filling in one gap only tends to reveal another question that could not be answered, or perhaps even conceived of being answered, before that was known. Palaeontology is no different, though when dealing with dinosaurs, the pieces filling in those gaps do tend to be rather large.

In the late 1700s and early 1800s, a series of palaeontological finds of giant reptiles across the south of England heralded the beginnings of a new understanding of the bygone Earth. These animals lived in ancient seas and were soon christened with a barrage of now familiar names – *Ichthyosaurus*, *Plesiosaurus* and *Pliosaurus*, and less familiar ones such as *Temnodontosaurus*, *Opthalomosaurus* and *Cryptocleidus*. Plenty of fossil animals had already been discovered at this point, but these were primarily those of well-known living groups of mammals like elephants and hyenas, or shelly fossils like ammonites, which had obvious relatives in living squid and cuttlefish.

But now there was inarguable evidence of major types of animal unknown in the modern world, and from a geological era where many familiar animals such as birds and mammals were apparently absent. These finds indicated that there had once been an Age of Reptiles, something quite unlike anything that scientific minds of the time might have imagined. This proved to be a sensation, with the learned public flocking to hear lectures on these amazing new animals from the scientists of the day.

This was a time of great growth of the natural sciences in Europe. Although Charles Darwin's grand theory of evolution by means of natural selection was still decades from publication, the ideas of species changing over time, and that species or entire groups could have gone extinct and were no longer alive, were under discussion in scientific circles. New discoveries in biology, chemistry and physics were fuelling new concepts about the world, and entire fields such as geology were being established. The hearts of the great continents in the Americas, Africa, Asia and Australia were being explored, and old fables were being banished as new information made it back to the learned societies of London, Paris, Berlin and others. It was a near perfect time to investigate whole new groups of extinct animals.

Before too long, great reptiles that had lived on land started to be found and recognised, in addition to those from the seas. Not for them the sleek shapes, paddle-like fins and tails of the ocean-going animals; instead they possessed more normal reptilian walking limbs, which pointed to a terrestrial lifestyle. Although these were initially known from only a few, very fragmentary pieces, but researchers quickly realised that they were an entirely new group of animals. They were christened with the name 'Dinosauria'. Despite this is commonly translated as meaning 'terrible lizard', a more accurate version is probably 'fearfully great reptile', which better captures the spirit of how these animals were perceived.

When it was published in 1859, Darwin's *On the Origin of Species* gave the naturalists of the time an evolutionary framework to understanding life on Earth both past and present. Indeed, this was a time when the fields of geology and palaeontology were very much in their infancy.* Science was all about discovering new phenomena, new species, new elements, and identifying physical laws, and despite the huge efforts in all of these areas, the scaffold for understanding the past was still, at best, very limited. Add in a healthy dose of biblical literalism – since many naturalists were trained by, or even ordained in, the Church – and these infant sciences can be forgiven much for their early errors and confusion.

---

* The two fields were in fact broadly synonymous at this point and known for a time by the delightful name of 'undergroundology'.

Even so, what stood out early on was twofold: that so much information could be derived from so little data, and that so much more remained to be resolved. This apparent paradox is to be a running theme of this book; people seem to be consistently amazed at what palaeontologists are able to work out about dinosaurs from the limited resources of the fossil record, while being equally amazed at things that are unknown.

The second dinosaur

The famous *Iguanodon* serves as an example of what could be elucidated at the time from very little. This was only the second dinosaur to be named (the first being *Megalosaurus*), the honour going to an English doctor named Gideon Mantell, who had become fascinated by all things fossiliferous in the south of England. Although the *Megalosaurus* was originally known only from a small number of somewhat leaf-shaped teeth, these alone were enough for Mantell to work out quite a bit about his animal.

First off, the sheer size of these – some were several centimetres long – meant that they must have come from a large animal. Second, they were almost certainly from a reptile, given both the serrations to the edges (very common in reptiles, and almost unknown in mammals) and the fact that they were from a time known to be dominated by reptiles and devoid of large mammals. The teeth also had long roots, implying that they sat inside sockets in the jaws. This feature separated them from most other reptiles (though is seen in crocodiles), where the teeth are all but stuck to the jawbones and lack roots, but this aspect seems to have initially been overlooked.

Finally, the overall shape of the teeth, and especially the nature of the serrations, were very similar to those of various herbivorous lizards alive today. In particular, these were near identical to the modern iguanas, thus the origins of the name – *Iguanodon,* meaning 'iguana tooth'.* The wear on the teeth showed that the animal probably ate

---

* Mantell originally wanted to call his animal *Iguanosaurus,* but was dissuaded on the grounds that this was too much like calling it iguana-reptile, which is rather repetitive.

tough plants and these, and indeed large herbivores generally, are rare in aquatic systems. Collectively then, from only a few teeth, Mantell was able to work out that he had the remains of a very large herbivorous reptile, which lived on land, ate tough plants, and was like a lizard but also somewhat different. It was also dissimilar enough from other known species at the time to give it a new name, and so in 1825 he published this as: *Iguanodon atherfieldensis*.

That's really a lot of information from a few teeth and shows the kind of inductive work and comparative anatomy that still stands as part of the basic toolkit of palaeontologists today. Still, it left more than a bit to be resolved, with huge uncertainties over this creature's size and proportions. As to what its head looked like, the only thing they had to go on was its teeth, and there was virtually no real information about such things as its skin or colour.

Soon though, much of a skeleton was discovered, and what later became known as the 'Maidstone slab' or 'Mantell piece' made its way to him. Now the good doctor had more material to work with, and early descriptions of some other giant terrestrial reptiles were starting to appear, allowing for some comparisons and generalisations about them to be made. Most of these animals would eventually be identified as dinosaurs, but that term had yet to be coined, and it was not yet clear if these animals were truly distinct from, for example, various fossil crocodiles.

Iguanodon was indeed a large animal with robust and strong bones. The shape of the femur (thigh bone) was straight and demonstrated that the leg was held vertically under the body, giving it an upright posture like a bird or mammal, and not out to the side with a sprawling posture, such as a lizard or salamander. From this, Mantell inferred that these animals may have been quick, active and agile, an idea that was controversial at the time, but that turned out to be remarkably accurate from so little information.

Already, though, some details were creeping in that, with the wonderful clarity of hindsight, turned out to be in error. Mantell and his peers were sufficiently able anatomists that they could put a disarticulated skeleton back together and make some reasonable guesses about the form of missing pieces, so it's not like there were arms mixed with legs, or tails were put together backwards. However, in his sketch of how the animal may have looked, Mantell had the hips and

shoulders, while in the right places, at the wrong angles. He reconstructed his new beast as a huge and squat quadruped, and an isolated
spur of bone found with the specimen was suggested as a spike on the
nose, giving the large animal a rhinoceros-like appearance.

This last issue is commonly cited to highlight the mistakes of the early
palaeontologists, the accusation being made that they were indulging in
some extravagant guesswork when they should not have been. However,
this misses a couple of vital points about the work being done at this
time and how people like Mantell were drawing on the limited available
information; not only from the few fossils they had, but also from the rest
of the natural world, which was still being uncovered.

Dinosaurs were different in various ways to the reptiles that came
before them and the living birds, mammals and lizards to which various researchers would have been able to compare them. There were
always going to be some unique features that would cause confusion
and, lacking any other even vaguely complete dinosaurs for comparison, it was inevitable that unique traits would be hard to interpret.
Context matters enormously. These early works were the first attempts
to describe some truly new animals. Given that there were so few of
them, and not an enormous pile of reference works available on other
species, errors were predictable, and indeed credit must be given to
the scientists, working as they were with such little information.

The second point that is overlooked, especially when it comes to
the nose horn, is that Mantell was doing something entirely sensible.
He wasn't comparing the larger and robust herbivore to a rhino
directly, but to the iguanas. Many of them have bosses of bone on the
nose, and one, the aptly named rhinoceros iguana, even has a pair of
them stuck one behind the other. Mantell was well aware of this; he
even included a sketch of the skull of one in a paper he wrote in 1841
and made the comparison rather explicitly. Lacking the evidence for
large bipedal reptiles and stuck with an incomplete skeleton, it was
entirely reasonable to propose a fully quadrupedal animal with such
an adornment on the nose.

All in all, it was a brilliant start, but there was much more to come.
Specimens of *Iguanodon* and other large terrestrial reptiles continued
to accumulate, and scientific descriptions of the new teeth and bones
appeared, allowing other researchers to add their input.

In 1842, Richard Owen, a legendary anatomist and the man who would later found the Natural History Museum in London, coined the name 'Dinosauria'. It was quite some claim to suggest that there was an important new group of reptiles out there, given that the dinosaurs at the time consisted of exactly three animals – *Iguanodon*, *Megalosaurus* and *Hylaeosaurs* – and none of them were known from especially complete remains – but time has shown that Owen was right to recognise that these were new and should all be grouped together.

Several other animals were known at the time that would later be recognised as dinosaurs, and plenty would soon be added from other discoveries. However, this small triumvirate were enough to show that these animals truly were different and special compared to the other finds of the time. *Iguanodon's* place in dinosaurian research was thus already assured, since it was the best represented of these newly recognised species. Looking back, it was a tremendous piece of insight from Owen to link these bits together as something special, but three fragmentary species, all from the south of England, would never provide sufficient information to say much about what dinosaurs were really like without much better specimens.

Happily, however, this problem was about to be greatly reduced thanks to a Belgian coal mine.

Skeletons by the dozen

In the year 1878, in the Walloonian town of Bernissart, a huge collection of dinosaur bones was discovered. Not only were there very large amounts of bones from a very large number of individuals, but complete and articulated specimens were unearthed, including the skulls. These animals were rapidly identified as belonging to *Iguanodon* (although it was given a new species name – *Iguanodon bernissartensis*), and thus between them and still more material that had been recovered in the UK, a new understanding of these animals was possible.

The nose horn was revealed to be a very unusual thumb, and presumably represented some kind of weapon. *Iguanodon's* arms were rather like its legs in general form, though shorter and more slender, suggesting the animal, even if it was a quadruped, was rather less elephantine (or even rhinoceros-like) in stature and proportions. The tail was not quite as long or lizard-like as assumed, and the head was certainly not that of an iguana, however large.

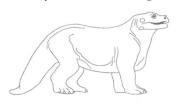

The changing face of *Iguanodon* over the years. Left, based on the model displayed in London in the late 1800s; middle, a typical 'kangaroo' pose common from most of the 1900s; and right, a modern interpretation. Illustration by Scott Hartman, with left illustration based on the work of Benjamin Waterhouse Hawkins, and middle that of Zdeněk Burian.

Also novel was both the number of specimens and the fact that they appeared to have died together in a massive group (this has turned out not to be the case, and the Bernissart dinosaurs most likely represent the deaths of animals alone or in small groups over many years, rather than one mass mortality). Such a find suggested large groups of animals living together, which would again make them different from most modern reptiles and pointed to more complex behaviours among these ancient creatures.

Thus our understanding of *Iguanodon* took a major step forwards. Not the lumbering beast of the earliest reconstructions with the rhinoceros-like horn and huge columnar limbs (as exemplified by the famous reconstructions led by Richard Owen in London's Crystal Palace), but a svelte and perhaps agile animal, that moved in herds. The discoveries pointed to creatures that were far from simply being large lizards, but a truly special set of animals.*

Dinosaurs everywhere

By this time the Americas were yielding their own great trove of dinosaurs. Whole new types of dinosaur were being uncovered across the Atlantic, and those known from only scraps in Europe were now represented by whole skeletons. At the dawn of the twentieth century, animals like *Stegosaurus*, *Allosaurus*, *Diplodocus* and *Triceratops* were well known in scientific circles and even to the public, and were the subject of huge debates and fractious discussions among palaeontologists. Dinosaurs appeared from further afield too at this time, with specimens now being found in Tanzania, India, Mongolia and Brazil (and with the imperialist attitudes of the time, these were shipped straight to Europe).

In places there was painfully little evidence available to help resolve the outstanding questions or come down on one side or other of a disagreement: for all this new knowledge and improved understanding of the dinosaurs, it was clear that there was a huge amount that we did not know about them. Were dinosaurs warm or cold blooded? Why did they die out? How did they get so big? How many different species were there? And how and why did they evolve into the plethora of forms that were already known?

Darwin's theories were now accepted science and new ideas about evolution, extinction and adaptation were settling into the mind of researchers. New fields such as ecology and ethology would shortly

---

* Some modern studies suggest that these two animals were not just different species, but actually different genera. Thanks to a quirk of taxonomic history, the Belgian animal retains the name *Iguanodon*, whereas the British creature Mantell worked on is now called *Mantellisaurus* in his honour.

arise (or become recognised as fields in their own right), giving greater context, and opening up new aspects and depths to our ignorance. while presenting new possibilities for understanding.

In this twenty-first century there are more specimens, more researchers, and more techniques and technologies available than ever before. Both our understanding and our ignorance have multiplied. It would be something of an embarrassment to palaeontologists, given the abundance of tools available to us now, were this not a golden age for research, but we are standing on the shoulders of the wealth of data previously accumulated, the power of analyses available to modern scientists, and indeed the huge amount of history that has come before.

Previous generations of researchers made plenty of mistakes, but science is self-correcting (eventually). Modern science allows us to learn from these mistakes and not make them again (hopefully). And, of course, palaeontologists of the past got a lot right and generated much of the vast amounts of data that we can use. Indeed, such is the reliance of palaeontology on original descriptions and details of specimens that it is one of the few fields in the sciences that makes regular and copious use of research published not just decades, but even centuries ago.

Dinosaurs now number well over a thousand species and are known from thousands of skeletons and many hundreds of thousands of less complete ones, along with bones, teeth and footprints, with their fossils recovered on every continent. There are specimens with scales, feathers, claws and even internal organs intact (or at least impressions of them), and eggs, nests and burrows have been found. Juvenile dinosaurs and embryos have been described and some tantalising claims of original biological material that have survived tens of millions of years, if not yet quite proven, are certainly credible.

With such information comes the possibilities that have not so much eluded science as seemed redundant. It was natural to assume that some dinosaurs were camouflaged and others brightly coloured, that some had spots and others stripes, while males and females may have been dramatically different in colour. Such features are all but universal among modern animals like birds, mammals and reptiles, so were assumed to be the case for dinosaurs. Yet with no possible way of determining the colours or patterns of these animals, the point was moot – it was not that we didn't know the details for the dinosaurs so much that we never *could* know, so there was little value in wasting effort speculating about it.

There were some reasoned extrapolations that creatures like *Triceratops* with its advertising billboard of a shield on its head might be brightly coloured, and that smaller dinosaurs living in forests would have disruptive patterns to help hide them, but that was about it. With no way to actually test these ideas though, they remained as reasonable, but ultimately unknowable, speculations.

Now, however, the spectacular preservation of soft tissues in feathered dinosaurs from China and Brazil, coupled with high-resolution imaging, have allowed traces of pigments and patterns to be discerned for a small number of animals. In one sense we have a revolutionary new understanding of some species, and yet for every dinosaur for which we know the colour, there are hundreds that we do not. An area of dinosaur biology that used to be considered virtually beyond our grasp is now ripe with possibility. We know today that we can potentially know something, but that we do not know it – a stark shift that emphasises what we do not, and may never, know.

Many more issues of this type are coming to the fore – areas that had been abandoned intellectually as being impossible to engage with, owing to a lack of data, are becoming rich seams of research and new ideas. As each is mined and examined, yet more information is revealed and the grand framework of our understanding of dinosaurs is fleshed out a little more. Even if it is a web of information, which is more hole than strand, the fundamentals are clear. What awaits is the gaps to be filled in and we are at a time when we are likely to see many of these completed.

We will start, however, with the end.

I

Extinction

WHAT ACTUALLY KILLED off the dinosaurs has understandably occupied a great many minds for a great deal of time. After all, how did, how *could*, such large and diverse animals vanish so abruptly? There has therefore been perhaps more thought (and also more speculative nonsense) devoted to this idea than any other in dinosaur research. Yet, as with so many other fields, we know an awful lot that is certainly not true, a great deal about what likely happened, and recognise too there is more to come.

There are literally dozens of ideas out there that have been advocated in print at one time or another for the extinction of the dinosaurs. Many are, by modern standards at least, unscientific or even downright absurd. To be fair, at least some of these were probably put forwards with tongue firmly in cheek, though others are equally implausible and yet were lobbied for by serious scientists with some apparent real conviction.

Dinosaurs did not die out because as a lineage they became collectively senile and forgot how to breed, nor were they hunted to extinction by alien invaders (and yes, both of those have actually been suggested). Plenty more theories can be ruled out quite simply because, while they might be plausible under the right circumstances, they do not explain the overall pattern of loss that happened at the end of the Cretaceous. The idea that mammals ate all their eggs could potentially explain the death of the dinosaurs (if very unlikely in reality), but it doesn't explain why crocodiles and turtles were not as badly affected, nor why the various marine reptiles that gave birth to live young also went extinct.

Other ideas are biologically unlikely, such as that a wave of diseases took out the dinosaurs. While obviously there are some nasty and dangerous infections in the world, and no doubt dinosaurs had their fair share, the suggestion that these could kill all dinosaurs everywhere more or less simultaneously (even on distant continents) is essentially impossible. Given the vast range of species involved and different environments inhabited, it's most unlikely that all dinosaurs everywhere would have suffered to the same degree; not least when very few infections have an exceedingly high mortality rate and there are almost none that we know of that are 100 per cent fatal in a single species, let alone across thousands. Again, the idea also fails to account for the loss of other lineages on land and in water, or the survival of birds, so can be safely rejected as a reason for a mass extinction.

Plenty of these ideas came thick and fast in the Victorian era, when people were first getting to grips with the idea that organisms could even go extinct. In the early days of geological sciences, a clear change in the types of fossils in various layers of rocks was used to help identify and separate out major periods in Earth's history. Geological formations in different places, both within countries and even between continents, could be aligned based on the sequences of various strata and in particular the fossils they contained. While most of these were done with the much more common marine shellfish such as ammonites and brachiopods, the dinosaurs and their kin had their place, too.

The 'Age of Reptiles' is not simply a colloquial term for the Mesozoic, but was in common use in the Victorian era as a formal name for this period. Rocks of this age could be easily identified because they would contain the bones of huge reptiles, most notably the dinosaurs on land, but also plesiosaurs and ichthyosaurs in the sea, and various other now long-gone reptiles such as pterosaurs and rhynchosaurs.

It is now firmly established that the Mesozoic Era ran from 252 to 66 million years ago and consisted of three major periods – the Triassic, Jurassic and Cretaceous (which were themselves subdivided into various categories). The relative positions were generally easy to establish, and while no accurate numbers might be in place, it was clear that the Late Jurassic came before the Early Cretaceous and that, within it, one could place certain fossils as coming before or after others. Each new find fitted the framework, and it was shown to be robust, with correlates from across continents soon to be discovered allowing us to understand, for example, that *Iguanodon*-being from the

Early Cretaceous – came after the Late Jurassic *Diplodocus*, and before the Late Cretaceous *Triceratops* of North America.

The obvious issue was that outside of these various geological formations that make up the Mesozoic, the great reptiles were no longer to be found. They were no longer alive to be preserved in the fossil record and therefore must have been extinct. To the Victorian naturalists and 'undergroundologists' of the time, extinction was a fairly new concept. It had been demonstrated for the first time only at the very end of the eighteenth century by the great French anatomist Baron Georges Cuvier.*

The French aristocrat was an influential scientist, who had a long interest in palaeontology, and was the first to describe a pterosaur, as well as having communicated with Mantell over the original *Iguanodon* teeth (in a rare error, Cuvier thought they might belong to a rhinoceros, something for which he later apologised to the doctor). In his treatise on elephant bones from France, Cuvier noted that they were distinct from the known Asian and African elephants and must both represent a different species – one that was no longer alive in France, or anywhere else. Thus, he concluded, at least some species must go extinct.

At the time, the community of academics looking at dinosaurs included a number of people we would now consider to have some form of creationist view – that the Bible was there to be interpreted at least somewhat literally and that the Earth and all things living and dead were created.

William Buckland (the first person to name a dinosaur – *Megalosaurus* – and later a canon in the Church) was among those prominent proponents of joining up these new fossil discoveries with religious texts and suggested that the great reptiles being found were those of the 'behemoth' and other biblical creatures.

However, even figures such as Buckland conceded that the dinosaurs were gone by the modern age, in contrast to modern-day creationists who bizarrely claim that dinosaurs still live in the Congo or

---

* He was famous enough for his deductive abilities on anatomy to be namechecked by Sir Arthur Conan Doyle in one of his Sherlock Holmes stories. Holmes notes that as Cuvier can reconstruct an animal accurately from the smallest of bones, he, similarly, reconstructs crimes from the most limited of evidence.

Amazon basins (or even less credibly, the US Midwest). Of course, aside from the birds (which are represented by more than ten thousand living species), the dinosaurs are clearly gone from the modern world and, despite numerous claims from fiction writers (and occasionally even scientists), they are not coming back.

There had to be *some* explanation for why these huge and powerful animals were no longer around, and plenty of scientists filled plenty of paper with their suggestions. These came and went as they fell in and out of vogue, or were simply impossible to prove. Numerous disasters were proposed at various times (cosmic radiation, massive floods, massive droughts, glaciation) to explain the end of the Age of Reptiles, but eventually one appeared that was both biologically plausible *and* had some evidence to support it. This was the suggestion in 1980 by the father-and-son team of Luis and Walter Alvarez that some kind of asteroid hit the Earth and resulted in a mass extinction.

The evidence presented was meagre to say the least: two bands of reddish and mineral-rich clay that immediately overlay the end of the Cretaceous and were rich in iridium. This is an element rare on Earth but very common in extraterrestrial rocks.

The logic was simple: a large interstellar body had hit the Earth, presumably relatively close to where the material was at its thickest, and scattered enough dust and particles into the atmosphere that bits of it could still be seen all over the world. That alone implied a huge body and a massive collision, one potentially big enough to create something of a nuclear winter where the sun was blocked out, chilling the planet and killing the plants, with the herbivores and carnivores inevitably following in turn. It would explain the global nature of the extinction and why so many disparate lineages died out and so quickly.

This hypothesis immediately caused something of a furore in geological and palaeontological circles, not least because the senior Alvarez was a physicist rather than a geologist (he won the Nobel Prize for Physics in 1968), as well as the suggestion that such a major rethink of a long-standing puzzle could be proposed based on two data points. Where, scientists somewhat reasonably asked, was the crater? A global killer would surely leave a sizeable hole in the Earth and one that might still be visible even 65 million years later.

The K-Pg boundary in Canada: here it is a layer of clay
that is several centimetres thick and is rich in iridium.
The scale bar is alongside this dark band, which has
bright orange flecks in it. Photo by Hans Larsson.

Decades of research have now vindicated the family Alvarez (and
their numerous collaborators). First of all, there are now multiple
sites known around the world where the red clay layer is known
and this shows a clear pattern of something that centred around
North America and was progressively less and less important further
and further away. In addition to the iridium, the mineral layer also
contains a number of other features associated with major impacts,
not least something called shocked quartz, which is formed only
under extremely high-energy collisions. More critically, we now
have a crater, and it is largely where you might expect it to be based
on the distribution of the iridium – the Yucatán Peninsula of south-
eastern Mexico.

The reason why it took a while to find was the fact that, perhaps
unsurprisingly given how much of the Earth is covered in water, the
asteroid hit on land but the rocks were marine in origin. The pieces
that were scattered around therefore gave the impression that it had
landed in the sea and, for a while at least, people were looking in the
wrong place. The crater itself collapsed and was filled back in, so it's

not exactly a nice clear circular deposit on the surface and is buried and largely inaccessible. But it's the right age, in the right place, with the right proportions of various minerals, and contains huge amounts of the shocked quartz that forms from such impacts.

All of these point to the landing of a major extraterrestrial body. With the crater known, we can also begin to piece together the likely size of the impact. An asteroid several times the mass of Mount Everest hit the planet travelling many times the speed of sound. The impact would have resulted in a transfer of energy in the realm of a billion times the force of a nuclear bomb. This is the kind of occasionally jaw-dropping figure in the sciences that is hard to put into perspective sufficiently for anyone to grasp, but, with all the British understatement I can muster, it was definitely quite big.

The effects of such an impact locally are also hard to conceive. There would have been an incredible earthquake, and the very air itself would have been set on fire by the pressure and heat generated by the blast. We can expect there to have been tidal waves and floods radiating out across the entire ocean, as well as potentially the triggering of multiple volcanic eruptions and other seismic events. For thousands of miles in every direction, affecting both land and sea, there would have been absolute devastation. One only has to look at the footage of events such as tsunamis and earthquakes to see how entire regions can be wrecked in minutes, but these effects would have been both considerably stronger in effect and massively more widespread.

Anything in the immediate vicinity of the impact site, and probably anything in a radius of hundreds of kilometres, would have been killed. Entire forests would have been flattened, coral reefs or other areas both in North America and well beyond would have been utterly annihilated. These effects alone would have made many species extinct. Where a dinosaur species lived in a localised area, or was reliant on some limited resource to survive, it would have perished, but that would hardly have made a huge difference globally. However, after this initial impact and local effect (if you can call a tsunami crashing on the far side of the Atlantic 'local'), the real long-term and global effects of a nuclear winter would start to kick in.

As a result of the dust and ash cloud blocking out the sun, the whole planet would cool, and rapidly too. Plants would start struggling or dying, meaning there was little or no food for herbivores of all kinds, and what was available would be of poor quality. With herbivore numbers dwindling, so too would those of carnivores. Add to this the altered climate, likely toxic water sources from ash and dust, and possible disrupted breeding seasons from the odd light and heat, and things would have become very difficult for organisms worldwide.

You don't have to wipe out every member of a species directly to doom it to extinction. Take out enough individuals of a population and they simply can't find a mate, or if they can, the gene pool is too shallow to adapt to changing conditions. Being in poor health might make them too weak to resist any diseases or parasites doing the rounds, or not healthy enough to undertake a critical migration or hibernation. They might not all be killed right away, but they can be doomed as a species, even if it may take years or perhaps decades for the last of them to go.

As more species go under or take a hit, so too will others. An insect might die off because it is too cold, but if it's the primary pollinator for a tree, then that species is also now only marking time until it goes extinct. The chain continues: anything that relies on that tree for food or shelter will become extinct eventually, and anything that ate those who sheltered in the tree are also now in terminal decline. In this way, the damage to only a few species or populations can ultimately trigger the collapse of entire ecosystems, and huge numbers of species, slowly or quickly, will go extinct.

Thus we now have an explanation in place for the death of the dinosaurs (and many other big reptiles). As large animals, they would have been especially vulnerable — larger animals have smaller populations, need more food and take longer to mature and breed than do small ones. When a crisis hits, there might be enough food left over for dozens or hundreds of rodent-sized animals, but not even one tyrannosaur. The biggest part of the biggest mystery of the dinosaurs seems to have been solved.

Or perhaps not.

Volcanic destruction

In April of 1815, the Tambora volcano in Indonesia erupted. It was one of the largest volcanic eruptions to have taken place in recorded history. Huge clouds of ash and dust were put into the atmosphere and what followed in 1816 became known as the 'year without summer'. Crops failed across Europe, North America and Asia, there were summer frosts in Europe, and southern Canada saw 30 cm of snow fall in June. Summers were cold in 1815, 1816 and 1817, and returned to normal only in 1818.

All of this was from one single volcano that exploded once, so imagine the effects of a whole chain of volcanoes that erupted sufficiently to produce a rock layer some two kilometres deep over an area of half a million square kilometres. These are the Deccan Traps of India and they represent the output of tens of thousands of years of eruptions across the subcontinent at the end of the Cretaceous.

As it happens, these eruptions occurred around a million years before the asteroid impact and one can easily imagine what the effects might have been, not least as they would have been strikingly similar to much of what has just been described. Dust, ash, varied temperatures, species suffering, ecosystem collapse. The asteroid impact almost certainly would have killed off the dinosaurs, but it may have been merely a timely cosmic *coup de grâce* for a great lineage already in terminal decline.

At various times a few studies have suggested that the dinosaurs may have been on their way out even before the asteroid struck. They point to low levels of diversity and reduced numbers of new species forming to suggest that there may have been some severe strains on the dinosaurs. This could certainly line up with the environmental pressures that might come from the Deccan eruptions. Still, the jury is very much out on this, as dinosaurs are not that numerous as fossils compared to many other animals. As a result, when there's an apparent fall in numbers, it is hard to determine if this is a real signal or a blip caused by the limited data.

Added to that, the variations seen in at least some numbers for the timing of the eruptions versus the loss of species makes them hard to reconcile accurately, and harder still to determine if the Deccan eruptions had anything other than (geologically speaking) a short-term effect. However, it does remain a strong possibility that, had the asteroid sailed past Earth without so much as a scratch, the dinosaurs would still have suffered enormous losses – and may have gone extinct regardless, if over a longer timeframe of slow depletion.

Survivors

While we have theoretical knowledge of the effects some combination of asteroid and Deccan Trap eruptions may well have had, what remains are some fascinating questions about exactly what occurred around, and especially after, the extinction. The most basic of these questions is why did all of the dinosaurs die out? As noted above, in some ways they were rather vulnerable in generally being large animals with long generation times; being small gives you a much better chance of survival.

Small animals need only limited resources to keep going (individually and for a viable population), so what will not even feed one elephant for a week or two could keep a dozen mice going for a year. They tend, therefore, to have much larger populations with more variation, and so a greater chance that some of them will have the genetic makeup to survive. They will breed faster, which can produce still more variation, and small animals can escape into microhabitats (like into burrows or hollow trees) that can help them avoid the worst of climate changes, especially keeping warm. So while the large dinosaurs were going to be vulnerable, there were also a good number of small lineages that should not have been at such risk.

Many were not only small, but seem to have been comparable to various species that *did* make it through the extinction event. In particular, there were numerous bird-like dinosaurs that were small and feathered, some of which could at least glide if not fly fully. Whatever traits that led to the survival of the birds might at least have allowed a few dinosaur groups to slip over the line and keep going. To be fair, the birds also took a hammering and many bird lineages went extinct at the end of the Cretaceous, though this merely opens up another question as to why some birds survived and others did not.

A 2018 study noted that the most vulnerable bird lineages were those that lived in trees, whereas those that were more terrestrial were the ones that survived. If anything, that suggests that smaller terrestrial dinosaurs (of which there were also plenty) might have been okay, so again, it seems odd that no small bird-like dinosaurs made it.

On a related note, various groups that we think of as being vulnerable to mass extinctions and especially climate change (most notably amphibians) seem to have got through comparatively unscathed, suggesting there were places that may have been relatively untouched by these cataclysms. If so, why didn't some dinosaurs persist there? As far as we know, dinosaurs were present and diverse in almost every terrestrial environment, so would they not also have been there in some tucked-away corner of the Earth?

This point belies a greater one: it really is unlikely that the dinosaurs (and again, so many other lineages) essentially went out in the geological blink of an eye, over a handful of generations and perhaps a few hundred, or even a few thousand, years. As shown by the survivorship of various lineages of mammals, birds, reptiles, amphibians and others, enough of the world was not so utterly devastated that various animals in various continents could not scratch together a living and survive, or even thrive.

There are numerous late surviving species and lineages alive today that are crawling on in low numbers or in isolated places (the tuatara in New Zealand, platypus in Australia, sloths in South America, lemurs on Madagascar, the coelacanth in the Indian ocean), despite their heyday having been and gone many millions of years earlier. These animals have persevered through many extinction events, so the survival of some dinosaurs would hardly rank as a great novelty.

Sure, there would have been tremendous pressure on these isolated ecosystems, but that hardly excludes the presence of the odd dinosaur here or there, and indeed some areas may have remained all but pristine. It's perfectly possible, even probable, that some large tracts of land continued largely unchanged for tens or hundreds of thousands of years after the extinction crisis with various dinosaurs doing just fine.

We have barely a fraction of a per cent of the world's surface available as bare rock for palaeontologists to explore, and of that tiny bit of area, we would need to have rocks that are not only from the right place where these dinosaurs may have survived, but also from the right narrow window in time, too. Thus, while it is perfectly possible that some dinosaurs survived for thousands, or even some millions of years beyond the extinction, it's probably all but impossible that we would ever find fossils from those sites.

Some recent studies have noted that the general situation in the aftermath of the extinction may actually have been more complex than we previously thought, but also perhaps not as hostile to life as imagined. Studying the life in the ocean is easier than on the land (where the fossil record is much poorer), and while it may be a poor proxy, it will at least give an idea of how things are going generally. Around the impact site itself, there was a functioning marine ecosystem within 30,000 years, and while that's obviously still a long time, it is very small in geological terms; if the impact site itself was recovering in tens of thousands of years, then perhaps others really were untouched to the point of allowing dinosaurs to (initially) survive.

Periodically, dinosaur fossils (generally teeth) do pop up in post-Cretaceous rocks and there is generally great excitement in the media (and among creationists) about the possibility of the non-extinction of the dinosaurs, and that there has been some terrible error by palaeontologists in assuming that the dinosaurs went extinct at all. However, we can be assured that they certainly did go extinct, though a few animals struggling over the line would hardly be a surprise, even if it would be very nice to find some evidence of this.

Unfortunately, to date, each example that has appeared has ultimately turned out to be a fossil that is out of place. Plant roots, rodent burrows, some re-deposition of eroded rocks, poor documentation, and the misreading of the geological layers can all lead to a fossil appearing to be in a place different to its origins. There remains the chance that one of these will eventually prove to be a true 'surviving' dinosaur, but for now we are left with a great unknown – did any dinosaurs actually survive one of the single greatest disasters to ever befall this planet? Certainly one can hope.

2

Origins and Relationships

WHILE THE EXTINCTION of the dinosaurs remains one of the most intriguing areas of dinosaur science, the origins of the dinosaurs are given far less attention, while being considerably more mysterious. It would be nice to start this chapter with something along the lines of 'Some 240 million years ago a new group of reptiles emerged, the dinosaurs', but if only we could be so confident. While we do have dinosaurs that were this old, it is hard to say how much further beyond this they actually went. Potentially at least, the earliest dinosaurs might have come tens of millions of years earlier, but it would take the discovery of a dinosaur of that age to show it. More simply, we don't know exactly how old the dinosaurs really are.

The timing of the origins of most groups of animals are generally difficult to pin down. Almost by definition, they would be a newly emerging lineage of organisms and so probably from a small population. They also quite likely had some evolutionary novelty that made them successful and so meant they evolved quickly or diversified rapidly. Put this together and that means that they were low in number and limited in space and time. It is not then a surprise that the fossil record tends to be very incomplete for such groups and indeed it is very impressive how many ancient groups we *can* trace back as well as we have.

All of the earliest dinosaurs we know of were small animals (1–2 metres long) and not especially common, making them both figuratively and literally rather small players in their respective ecosystems. These first dinosaurs were all bipeds, and either carnivores or omnivores – the bigger animals that walked on all fours and/or ate plants came only later. They had four or five fingers on their relatively long arms, and would have been fairly fast moving and agile compared to some of the other animals that they lived alongside. By the standards of later dinosaurs such as *Stegosaurus*, *Diplodocus* and *Tyrannosaurus*, they were all fairly uniform and uninteresting, in that they lacked obviously exciting features such as armoured plates, long necks or big bony crests.

The first dinosaur?

We know that the dinosaurs split off from an ancient group called the dinosauromorphs (that is, quite literally, 'dinosaur-shaped'). These, as you may imagine, look very similar to early dinosaurs, and with incomplete skeletons it can be hard to tell them apart; indeed plenty have been confused for early dinosaurs and vice versa at various times. These animals would have lived alongside each other, too, and although the dinosauromorphs were not around for that long and the dinosaurs eventually replaced them, there was plenty of overlap; for tens of millions of years, the two lived side by side in various ecosystems.

In 2013, a redescription of a small skeleton first found in the 1930s in Tanzania resulted in the naming of a new genus of reptile, *Nyasasaurus*. A number of animals close to the origins of dinosaurs and their immediate ancestors come from Tanzania and so this has been a hotbed of recent research for the origins of the dinosaurs as a result. The authors of the new study produced an analysis of the relationships of early dinosaurs and their ancestors and found that it was hard to tell if *Nyasasaurus* was actually a dinosaur, or the nearest relative of dinosaurs yet known.

One thing that is notable about this animal is that it has three vertebrae making up part of its hips. Most reptiles have only two and the dinosaurs all have more, so this does point to the potential of *Nyasasaurus* being a true dinosaur. Adding bones to the arrangement of the hips makes for a stronger foundation for moving around and supporting the body, and was a key development in the evolution of much larger dinosaurs later on. From the same fossil beds come some unusual bones from the neck of a so far unidentified animal, but which may belong to a very early carnivorous dinosaur, and so again the true Dinosauria may be around at this point.

So were the very first dinosaurs from East Africa? Perhaps, but in the last few years there have been a large number of finds of dinosauromorphs from South America, including Brazil and a place previously largely devoid of early dinosaurs and their relatives, Venezuela. There are also more to come, with ongoing excavations turning up good numbers of material that will likely produce yet more new species, or at least greatly increase our knowledge of those we already have.

These new discoveries come on the back of already known forms from South Africa, Scotland and the USA among others. These, if anything, add a raft of complexity to the question of exactly when and where dinosaurs first appeared – clearly there are plenty of good candidates, but they are not so far apart as one might immediately imagine. Back at this time, the end of the Middle Triassic, the world was effectively composed of a single giant continent called Pangea and it was much easier for animals to move around. A trot from North America to India, Australia, Spain or even Antarctica was quite feasible.

It would still be thousands of kilometres, but relatively free of mountain ranges or oceans, and so while a single animal might never make such a crossing, populations could clearly move long distances over the generations. A good example of this is the animal *Lystrosaurus*, an early branch of the lineage that would go on to produce the mammals. In the early Triassic this animal was extraordinarily common and its remains can be found in western Russia, China, South Africa and Antarctica among others.

Thus it is not a surprise that the dinosauromorphs are so well spread out and that very similar forms are found in, what are now, such distantly placed locations. Even so, it would be nice to pin down a likely original point for them, as this would allow us to look more closely in rocks of this area and perhaps get a better handle on their earliest days, and how they then spread further afield. This is something we can't easily do now.

Going so far back in time means that differences between the various lineages branching out at this point in evolutionary history are rather limited and it is easy for us to be unsure over the exact relationships between such groups. With incomplete specimens and gaps in our understanding, when we obviously don't know what we don't have, it could be there are better candidates in our collections already, but we simply don't know yet. *Nyasasaurus* sat around for the thick end of a century before it really got the attention it perhaps needed. As we shall see shortly, this uncertainly over relationships extends right into the dinosaurs themselves, and areas that we thought we understood well can still be overturned.

The rate of new discoveries of the dinosauromorphs (and other lineages from around this time that also help give us context to how various groups evolved) is such that we are likely to do much better in this area in the near future. Although the origins of groups can be difficult to ascertain in detail, it's also true that the more finds the better, and we have many times more now than we had only a few years ago. Also important is that we have both new species and often multiple examples of these animals. That means we have a pretty good idea of their entire skeletons, and any oddities in one part of the body (like convergence in tooth shape or hand structure) that could lead to us confusing one group for another when represented by only a few bones are minimised.

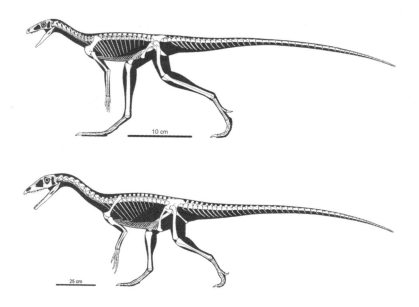

Drawing of the skeleton of a typical dinosauromorph, in this case *Marasuchus* (above) that can easily be mistaken for early dinosaurs such as *Tawa* (below). Both animals were small, with the former being around 50 cm in total length and the latter not much more than 1 m. Illustrations by Scott Hartman.

A huge amount of work is underway on these various species, with teams in Brazil, the USA and the UK working on them (and with various international collaborators too), but these are early days and it is hard to predict what the results may be. A single find that is maybe 10 million years older than the current record could produce a major shift in our understanding. Already *Nyasasaurus* has been suggested in one analysis to be a dinosaur, but not an especially early one in evolutionary terms. If that is the case, it would at least suggest an even earlier origin for the group, for them to have had time to diversify and evolve. So even based on our current knowledge, the dinosaurs could be rather older that we can currently demonstrate.

Getting started

Why the dinosaurs evolved at all, and how they came to take over, remain two areas of major contention. The Triassic was a time of great diversification of animals, especially the amniotes (that is, the group that consists of the reptiles and mammals and their ancestors). At the end of the preceding geological era, the Permian, there was the greatest mass extinction the Earth has known.* Estimates vary, but approximately 50 per cent of all species on land and 95 per cent of those in the ocean perished. As a result, those lineages that survived into the Triassic were in environments that were largely free from competition. Normally we would expect slow progression and changes through natural selection and other evolutionary mechanisms, but with few competitors, novel anatomical or behavioural features can flourish. New groups, therefore, both appeared and diversified rapidly (and by extension left relatively little fossil evidence) and while the ancestors of mammals were still on the scene, it was the reptiles that came to dominate.

---

* Through geological history there have been five mass extinctions where at least 50 per cent of all species alive at the time have been wiped out (and many smaller extinctions besides). The exact cause of the end Permian extinction is still uncertain, but something like 95 per cent of marine species and 75 per cent of those living on land died out. Since the origins of the first cells, this is the closest that life has come to being wiped out on Earth.

Among these new reptile groups was one called the Archosauria (the 'ruling reptiles'), which included the crocodiles, pterosaurs and the dinosaurs, as well as various other relatives (including, of course, the dinosauromorphs). In general, this group had one feature that appears to have given them an advantage over their other reptilian rivals – an upright stance. Modern reptiles walk with a sprawled posture, where the legs are held out to the sides of the body, which contrasts with the upright posture of birds and mammals, where the legs are held under the body.

When extant reptiles walk, the whole body tends to swing from side to side and the arms and legs move around in an arc, but in the upright animals the limbs are under the body and move parallel to each other and the direction of travel.* This is a much more efficient way of moving and generally allows such animals to be faster, too, so it is easy to see how the archosaurs generally might have been at an advantage over the average reptile of the time.

What is not clear, though, is why of the many archosaur lineages around in the Triassic, it was the dinosaurs that came to dominate. At the end of the Triassic, a second mass extinction hit the Earth. Probably caused by mass volcanic eruptions and the release of huge amounts of carbon dioxide, it was nothing like as bad as that which came at the end of the Permian, but it still did some terrific damage. Many of the new reptile lineages went extinct or were hit hard and never again attained their previous levels of diversity. The dinosaurs not only survived but flourished, and dominated the terrestrial ecosystems for the next 135 million years.

They may have had an advantage over the non-archosaur reptiles, but there is nothing obvious to separate them from their fellow dinosauromorphs or the other archosaurs. Although the pterosaurs were in

---

* Readers will have noted that crocodilians are above given as archosaurs, but have a sprawling gait. That is because ancestrally they were indeed upright and fully terrestrial animals and there are many such examples in the fossil record. Some later crocodilians became semi-aquatic and as a result switched back to a more sprawling posture, as this allows the limbs to fold alongside the body when swimming and makes them more hydrodynamic. These were the ancestors of today's modern crocs and their relatives, and hence the difference. In fact, even modern crocs can carry out a 'high walk', where they haul themselves up into a more upright posture than the lizards can manage, but it's a bit of a compromised way of moving.

the skies, plenty of crocodile lineages were essentially terrestrial and there were other archosaurs around (including many that were much larger than the average early dinosaur), such as the huge-headed predators called the rauisuchians.

Quite what prompted this initial survival that led the dinosaurs to survive the Triassic extinction and be left with a largely free planet (or at least devoid of large competitors) is difficult to say. Studies of the evolutionary rates and diversification of the dinosaurs versus other archosaurs shows that the groups were generally well matched. The dinosaurs were not evolving faster, gaining new traits, or diversifying in ways that might have given them an advantage, either in direct ecological competition for resources, or a general one of having more and different species that would increase the odds of them having one lineage or another surviving to take over.

The obvious conclusion is that dinosaurs got lucky. They simply happened to have the right species in the right place at the right time, such that when the extinction came, some of them made it through whereas other groups did not. It is more than possible that there is a more complex explanation than this, but certainly species and lineages can survive unexpectedly, and those that seem well suited to get through extinctions are sometimes wiped out. It is a rather unsatisfying explanation, but the best we have for now, though as ever there's plenty of room for new data and new hypotheses to provide a more compelling narrative for the early survival and success of the dinosaurs.

Part of this puzzle comes from the fact that we don't really know what caused the end Triassic extinction. Unlike the extinction of the dinosaurs, where it's pretty obvious that the asteroid did it (whatever effect the Deccan traps might have had beforehand), there's no clear answer on this one. Multiple ideas have been proposed and, while plausible, all of them have issues. There may have been another extraterrestrial impact, or a round of major volcanic eruptions, or the oceans may have become acidified (a quite common effect of volcanic eruptions) coupled with changing sea levels and thus changing climate.

In particular, recently the so-called CAMP (Central Atlantic Magmatic Province) volcanic events have been favoured as the source of this extinction and are currently the most likely candidate to win out as being responsible for the turnover of life at the end of the Triassic. Huge eruptions similar to those described for the Deccan traps have been shown to have started right around the time of the extinction, placing them squarely in the frame. Still, it hardly clears up why the dinosaurs got through this okay, when so many other archosaurs and reptiles did not.

Whatever the cause of the end Triassic extinction, it was the demise of many of the archosaurs and the start of the reign of the dinosaurs proper. Even so, at least a few dinosaur lineages were on the way to a fair amount of diversity. Although the earliest ones were very much in the mould described above, even before the extinction the first sauropods (the long-necked giants that include animals such as *Diplodocus* and *Brachiosaurus* in their ranks) were on the scene. That alone is interesting, as whatever else happened for the dinosaurs to survive into the Jurassic, it was clearly not only one lineage that made it in the form of the sauropods, but at least a few.

Critical to understanding almost any major aspect of the dinosaurs are their evolutionary relationships – who is related to whom, how closely and what features they inherited from their ancestors, and what is novel or special to them. At the most fundamental level, the dinosaurs have long been split into three main groups: the theropods, sauropodomorphs and the ornithischians.

The carnivores

The theropods are exclusively bipedal and include essentially all of the carnivorous dinosaurs. Although various members of the theropods were later switched to being herbivorous, if it was a carnivore, it was a theropod. There are some possible exceptions of a few of the earliest dinosaurs from other groups, where these may have been carnivores or omnivores too, but early on both the sauropodomorphs and ornithischians adopted a herbivorous lifestyle and stuck with it ever after.

The largest theropods were in the range of 12–14 metres long and around 6–8 tons, with the smallest about the size of a chicken. There was a variety of sharpened teeth and claws on offer and plenty of theropods had a beak; lots of them also had feathers. Famous theropods include *Tyrannosaurus*, *Allosaurus*, *Velociraptor*, *Spinosaurus* and *Archaeopteryx*, and the birds are ultimately descended from this group.

The giants

The sauropodomorphs are most famous for their huge size and the long necks and tails that were mounted on barrel-shaped bodies, and they walked on all fours. Although this body plan will be very familiar and everyone will recognise names like *Brontosaurus* and *Brachiosaurus*, these animals are the sauropods. As you may realise, this means that there is also another form of sauropodomorph, the so-called 'prosauropods' (i.e., those that came before the sauropods).

The earlier members of this group that abound in the Triassic had some of the features of their later cousins, but in a somewhat different format. The prosauropods did have long necks and long tails, but they were bipeds. Limited to two legs, they could not spread the weight around in the way that was possible with a quadruped and so never got to the sizes of their more famous relatives. The prosauropods were, however, still big compared to the other terrestrial reptiles around at the time, as the largest of them were around a ton at full size; though that's a fraction of the largest sauropods, who reached at least 50 tons and perhaps as much as 80.

Together, the theropods and sauropodomorphs are grouped together as the saurischians – a name that effectively means 'lizard hipped'. They were not sprawlers like lizards, but a key bone that makes up part of the pelvis, the pubis, had an extension that pointed forwards towards the head, as in lizards and crocodiles. This stands in contrast to our final big group – the ornithischians or bird. In these, the pubis had an extension that pointed backwards, as with modern birds.

An obvious issue here is that birds being descended from theropods are a member of the lizard-hipped group and not the bird-hipped one. This is because a number of the later theropods evolved to have a backwards-pointing pubis and so independently acquired this feature. At the time the various groups were named in the 1800s, there were too few clues that birds might be related to dinosaurs, and so the saurischian and ornithischian names were reasonable descriptions of the anatomy at hand, and not intended to imply evolutionary relationships with the reptiles or birds directly. It's an annoying quirk of history and evolution that such a position has come up, but there we are.

The weird herbivores

Turning to the ornithischians, again aside from the earliest forms that may have included some omnivores, this group are exclusively herbivorous. They essentially include everything not in the sauropodomorphs or theropods, and if the creature was a bit odd-looking and ate plants, odds are you had an ornithischian, especially if it had some kind of elaborate armour or bony plates on its body or crest on its head. Thus, such familiar and diverse animals as *Stegosaurus*, *Ankylosaurus* and *Triceratops* are all ornithischians and, of course, have a backwards-pointing pubis.

Although these animals never reached the size of the sauropods, some of them were very large (*Ankylosaurus* might have reached 10 metres long and the biggest hadrosaurs were probably over 25 tons) and in terms of diversity, they produced more variety of forms than the saurischians did, with various bipeds and quadrupeds, armoured forms, those with huge domes of bone on their heads or horns, and at least one species that burrowed.

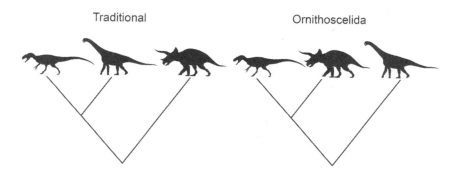

Competing hypotheses for the relationships of the main
dinosaur groups. On the left, the traditional model
with the theropods (represented by *Allosaurus*) and the
sauropodomorphs (*Camarasaurus*) joined together to form
the Saurischia and they are joined with the Ornithischia
(*Triceratops*). On the right, the alternate proposal with the
theropods and ornithischians making up a group termed
Ornithoscelida, and then the sauropodomorphs.

Family Trees

These three major groups (the theropods, sauropodomorphs and orni-thischians), or more formally, clades, are made up of dozens of different subdivisions that palaeontologists have identified, which give names to various groups of species that are closely related to each other. In much the same way that we recognise a large group of (mostly) carnivorous mammals as being carnivorans, made up of groups like cats, dogs, bears, seals and weasels, and then within dogs there are groups such as the foxes and wolves, so too can dinosaurs be divided and subdivided. It is a convenience (attaching names to them rather than numbers, for instance), but these clades also mean something; they are defined by the evolutionary history of the groups that they represent.

Working out what these relationships are and in doing so recon-structing a family tree of dinosaur evolution has taken a lot of work and remains not without controversy, though most researchers would agree on the general pattern shown here. The few outstanding issues are ones that are likely to be resolved without too much stress and won't change the fundamental patterns.

For example, it's not entirely clear if there really is a true evolution-ary clade of relatives of the huge sauropod *Brachiosaurus* (called, as you may expect, the brachiosaurs). Certainly there are a number of species that are similar to each other and share the basic features of especially long forelimbs and some details of the vertebrae that make up the neck, but whether or not this represents a major branch of sauropod evolution worthy of a name, or merely a few species that appeared along the evolutionary highway of one group becoming another, we don't know.

To a degree it may not matter; those species did exist and we know which other sauropods they are related to, so while it is an itch worth scratching, however it resolves (or even if we can't easily resolve it), it won't change our general understanding of sauropod relationships.

A more taxing issue is exactly which group of theropods is closest to the birds. Pretty much every analysis of the last ten or even twenty years shows that this is one of two groups of small feathered animals called the dromaeosaurs and troodontids, so there's no argument that it is probably one of them, but here things break down. Some analyses favour one, some the other, and many cannot pick between them, and each comes out as being equally likely.

Worse, in some recent studies, *Archaeopteryx*, universally regarded for 150 years as 'the first bird', comes out not as a bird, but as part of a new group, meaning there would be three groups now competing to be the nearest relatives of birds, not two. This is similarly one that doesn't exactly challenge our understanding of the relationships of dinosaurs to birds (or even these groups to birds), but it does mean that we can't establish a clear sibling group to birds and therefore know which group to concentrate on when it comes to understanding the origins of flight. That both troodontids and dromaeosaurs seem to have had gliding animals in their ranks also helps to confound the question of how birds became airborne, a subject we will return to in Chapter 15.

Ultimately this comes down to the lack of data to clear up these differences. Very old species are often incomplete and in the case of these dinosaurs, while the specimens are often spectacular with their preserved feathers, they are also crushed flat, meaning details of the bones we rely on are obscured.

In the early days of taxonomy, things were rather more simple. A biologist would look for a few key features shared by various species or lineages and use them to work out who was generally related to whom. We do this ourselves almost instinctively, and mostly accurately, all the time. You don't have to be an expert in reptiles to spot that while different, alligators and crocodiles have more in common with each other than almost any other living thing, and they are also more like lizards and snakes than they are mammals or beetles. At a finer level, you might want to look at the nature of their teeth or structure of the scales, but a few gross traits should set you well on course. This is generally fine, but it also breaks down quite quickly.

Things like sponges or bacteria don't have many anatomical traits you can compare easily to other species within the group or between groups. It might be easy to lump the crocodiles and the alligators together, but what about the other crocodilians such as caiman and gharials? And within the crocodiles, which species are most closely related to which others, and who is more distant? Where exactly do they go with the lizards, before or after, or even within as an odd branch of the lizards? These questions can't be answered by a few choice features (and indeed, which features should you pick, and are my features better or worse than yours, and how would we know?).

Step forward German entomologist Willi Hennig, who wrote a book in the 1960s proposing a new and formal way of looking at evolutionary relationships. Essentially, he advocated that rather than looking at a few key traits, biologists should look at as many as possible. Brains, teeth, limbs, skin, colours, numbers of chromosomes – ideally hundreds. Limited initially by computing power, it soon became possible to compare hundreds of traits between dozens of species or lineages, and developments in genetics technology meant that soon genes or even individual DNA bases could be compared.

This was a huge leap forward in approach and the depths of the analyses available. Numerous problems in taxonomy and evolutionary relationships were resolved (although it did generate some others), but perhaps understandably, by including more data there was a much better idea of what went where.

This also helped avoid problems that occurred when two groups had independently evolved similar traits. There used to be a group of mammals recognised called the Pachydermata, which consisted of elephants, hippos and rhinos. As you may imagine, this came from their unifying feature of thick skin as well as being large herbivores. But these features have all come about independently (in evolutionary terms we would say they are convergent) and a look at the details of the anatomy of these animals (and now too their genes) reveals it's clear they are from three separate lineages. Hippos belong with the antelope and deer (and, as it happens, whales), rhinos with horses and tapirs, and the elephants are a separate group rather more distantly related to any of these others.

However, such a trend for data collection is a mixed blessing for palaeontologists. We simply can't acquire data on the genetics of extinct animals, or their general physiology or behaviour, and so we are left with only the anatomy of hard things like teeth and bones. Even here, most dinosaurs are incomplete or problematic – a specimen may be missing arms or legs, or be little more than a few teeth. The bone might come from a juvenile, which doesn't necessarily look like the adults of other species you want to compare it to, and bones can be crushed or distorted or generally damaged. So we are reliant on using one tiny part of the whole animal (the skeleton and teeth), but even that can be missing the vast amount of the data you could in theory record and compare.

In fact, it should be celebrated just how much we can get from such incomplete and problematic remains, but it does mean there are huge gaps in our understanding. A specimen known only from a skull can't really be compared to another known only from limbs, and a third known only from the tail. They might all look like they come from the same species, but without overlap in the bones, it would be impossible to say. Thus discoveries of new specimens are always valuable – even species that are well known might have gaps – and understanding areas like variation (not everyone is the same height or build, etc.) is important.

This is the problem we have with working out the key relationships at such branching points, both for the origins of birds and, as mentioned before, for those of the very first dinosaurs. Here we have species and lineages beginning to separate from one another that are a long way back in the past. The further back in time you are searching, then, on average, the fewer fossils you have, as they have had more time to be lost or destroyed by the Earth's processes of erosion and subduction. Trying to reconstruct how they relate to one another, based on only a few, incomplete and very similar specimens, is therefore hard.

That is also why there may have been various omnivores at the base of groups otherwise considered to be herbivores: they may not yet have evolved into being herbivores, or we have misplaced a small carnivore into the wrong lineage!

Mixing up the dinosaurs

And so from here we turn to Ornithoscleida. This was a proposed linking of the theropods and the ornithischians to the exclusion of the sauropodomorphs; in other words, something completely against the 'traditional' saurischian–ornithischian split. This was first proposed back in 1869 by the legendary anatomist Thomas Henry Huxley, but it never gained much traction. Sauropods (the early prosauropods were barely known at this point) and theropods clearly had a number of features in common, most notably the structure of the pelvis and the fact that both had pneumatic, bird-like bones with extensions of the lungs filling them, making them hollow.

The ornithischians obviously lacked these and also had some extra bones in the skull that neither other group possessed, which suggested that they were different. So the ornithischian–saurischian split remained, not least when dozens if not hundreds of analyses since then also recovered the same fundamental relationships, other than a few early species jumping between groups.

Ornithoscleida as an idea was therefore essentially forgotten until 2017, when a major analysis was undertaken by British researchers looking at the earliest branches of the dinosaurs. What came as rather a shock was that they failed to recover the traditional relationship and instead found the theropods coming out with the ornithischians. This was in palaeontological circles quite shocking. It was not so much the change itself or the overturning of a long-held idea (if the data and the analysis are good, then that's the answer), but that this was so unexpected. Many big changes have occurred in our understanding of dinosaur relationships over the years, but they usually had some major trails attached to them.

There were ideas around that had come back more than once because of new lines of evidence, or that were always somewhat uncertain, or vacillated between several options. So when convincing evidence of a change came in these it was not much of a surprise – it had been on the cards, or at least registered as a possibility. But here the standard ornithischian–saurischian split had essentially been unchallenged and not even considered as being potentially incorrect by the vast majority of researchers. Sure, there were new studies coming out looking at the base of the dinosaur tree, but there was no particular indication that this was likely to shake things up so much as add to the usual slight round-robin, or settling out of some problematic species.

Such an idea was always likely to stimulate huge new interest and so it did. Very quickly another team of researchers had a look at the same issues and published a revised and different conclusion. They didn't exactly support the original result, but neither did they recover the expected result and their analysis was unable to pick between them. At the very least then, it seems that the fundamental relationship among dinosaurs is up for grabs. Oddly enough, the details are unlikely to change here – the tips of the branches of the family tree of dinosaurs are stable; it's more a question of working out which branch comes out where.

What this does is leave us with a truly burning question: how do these lineages relate to each other, and by extension, what evolutionary pressures caused this split? If we can work out the correct answer (assuming that we do not go back to the original ornithischian–saurischian split), then we can start to look at how various traits have evolved and what this might mean for the very origin of dinosaurs. Right now, the jury is not so much out, as still being presented with evidence, with more accumulating and the detail and level of analysis still growing.

This is an area of such importance and interest that it will attract a huge amount of research in the coming years and will no doubt be resolved fairly soon. How it will resolve is perhaps anyone's guess, and given that the last time we thought we had this fixed it suddenly changed on us a century later, it might be some time before we are confident we really do have it nailed down.

We may assume that a given group of dinosaurs had gone extinct by a certain time because there is simply no evidence of them after this point, but one may pop up later and show this was not the case. All the new specimens and reanalysis of old ones coming out around the origins of dinosaurs is one perfect example. That doesn't mean that our foundational understanding is flawed, though. The basics hold and continue to be supported, and it is notable that even a major shift like the possibility of an ornithoscleidan grouping is not such a big issue, given the known gaps in our understanding, and this had already been proposed based on solid evidence at the time. It's more a reversion than an overturning of our information.

3

Preservation

D ATA ON DINOSAURS is fundamentally limited, because the fossils are usually only skeletons and missing most of the living animal. But on top of that, it can be very hard to extract more than the more obvious surface details from them. You can easily see the shapes of bones and teeth and look for details on their outsides, but getting into the bone is another matter.

One solution is to cut them up and look at the insides, either for gross detail or to look at pieces under a microscope. Such destructive sampling is often prohibited or limited, since there is no guarantee palaeontologists will uncover any other specimens of a given species in the future, so you don't want to ruin the only example you might ever have. And, of course, any kinds of comparison between males and females are impossible with only a single specimen.

All of this makes it hard to work on fossils, but even this is not the end of it. Most specimens don't have even half the bones of a skeleton, and one or two hundred million years underground can leave them in less than pristine condition. They will usually have undergone some form of damage, distortion, erosion, chemical alteration or other effects, meaning that they are not even perfect representations of the bones that they once were. So even the data that we *do* have is less than ideal. Understanding how and why those changes happen and what that means for the fossil in hand is therefore critical to making the most of it, and being able to interpret that limited data correctly.

If you want to study the biology of any living species of animal, then it should not be too difficult to get hold of one. Sure, some are small and hard to see, others are rare and hard to find, and plenty live in inaccessible places like the deep oceans or up in mountains, but with some time and resources it should be possible to monitor them while they are alive, be it in the lab (or zoo), or in the wild. If you are happy to work from dead specimens, then things become even easier – museums will have collections, as will many universities, and it would not be too hard to travel to see a good selection of any given species you had in mind. Data is available, and provided that there has not been too much decay, it should be about as complete as it can be.

The contrast with dinosaurs is obviously stark. Most species are known from a single specimen and then only from bones and teeth. Soft tissues like skin, muscles and viscera are even more rare; data such as the genetic code of the animal or information on their proteins is unknown; and their behaviour and ecology in the wild can hardly be studied. Even getting to the limited material that there is can be a real pain – with only one specimen potentially available, it means that researchers can't simply travel to the nearest large museum or university, but might have to go halfway round the world to see it.

Even if you get there, getting data out is rarely trivial. Fossilised bones do not scan easily as they are very dense, and while X-rays, CT scans and others can be applied to them, they are often unsuitable for scanning systems that are usually there to examine differences between muscles and bones, not rocks and fossilised bones (i.e., other rocks). Plenty of dinosaur fossils are also some combination of too fragile, heavy, large and valuable to be moved to a location where they can be scanned and assessed.

The study of the process from the death of an organism to the recovery of the specimen is called taphonomy and it is a key aspect of palaeontology. All fossils have been through some fundamental process of transformation from (dead) organic tissues to lithified minerals – they have fossilised. At various times, however, they can have undergone any one or more of numerous processes and effects that can profoundly change their nature. Coupled with some major issues about which animals are likely to be preserved, and how environments and habitats can influence that condition, and there is a whole history to almost any given fossil that potentially can mislead or misinform.

Palaeontologists have for generations recognised the underlying problems at play with regards to issues such as erosion and distortion of fossils, but recently there has been a great expansion of experimental taphonomy – testing ideas in the lab and field about decay and preservation – which has expanded our understanding enormously. Still, there is much more to come in this area and there are enough problems that we do know about, and cannot immediately resolve, before we move into the future.

Becoming a scientific specimen

What determines whether or not an animal will become a fossil in a collection? This is at one level a simple enough question – it must be buried in some form of sediment, avoid decay, and instead undergo the chemical changes that will turn it from bone to rock, survive underground for perhaps millions of years, become eroded out of the surrounding matrix, be discovered and then collected and protected. This simple sequence hides a great many details and variables and also rather skips over what makes the difference to being buried or not in the first place.

Dead things decay. But different parts decompose at different rates and there's a good reason for the Hollywood cliché of tombs full of skeletons. Bones are living tissues but they do dry out once their owners have passed on and there is an awful lot of mineral there. So while skin and muscle and viscera will tend to disintegrate quickly, feathers, claws and ligaments will take rather longer, and bones stay around, too, although even skeletons will fall apart eventually. This decay is mostly coming from bacteria and fungi, though mechanical actions of water and soils can also hasten the breakdown. It's also likely in most circumstances that larger scavengers from beetles up to other dinosaurs might have got in on the action.

Enough of this can occur to utterly destroy the body of even a large animal, such that little more than fragments of bone remain. If a fossil is going to be formed, then at some point the animal needs to be buried before it has gone this way, and the sooner the better.

The environment plays a huge role in this. Decay will be faster in hot and humid climates like rainforests and tend to be slower if it's very cold (such as around ice), or very dry (deserts and some caves). Other conditions hostile to life such as in the acidic and anoxic (lacking oxygen) waters of a peat bog can also all but stop bacterial breakdown, and tend to house fewer animals generally that might try and make a meal of a recent corpse.

This means that we end up with some rather odd inversions when it comes to fossils – animals tend to be rare in deserts, but are in ideal conditions there to avoid decay and are likely to end up fossilised, whereas tropical rainforests are hosts to huge numbers of species, but decay is so fast they rarely produce fossils. We have an excellent idea of what lived in desert conditions, but almost no knowledge of those places that would normally host the greatest diversity. It's a quite incredible reversal of fortunes and one that obviously skews the data that we have and how we may perceive ancient worlds based on the fossils we find now.

Moreover, things also need to be buried. If you die in the middle of a forest floor or up in the mountains, there's not much chance of you being covered up by a good dose of sand, silt or mud that would stop you decaying any further. If you are on a floodplain, however, the odds are good, and if you fall into a river or lake they are even better. Deserts might not have much water flowing, but they do tend to have lots of sand moving around on the wind and entire sand dunes can collapse, which also works well.

Ideally, you will also not move much. A body can float downstream and even out to sea, losing bits on the way and generally getting bashed and beaten up as well as exposing it to more possible scavengers. The sooner you can be buried and the less accessible you are to anything that might break you down further, the more of you might become a fossil.

The more familiar and more common form of fossils are generally decent representations of the original bones. They often suffer some form of compression over time (more on that shortly), but they are basically three-dimensional representations of the originals. Occasionally, bones can chemically dissolve or disintegrate even as the material around them is compressed into rocks. In these cases, the bones might be missing but the space can be retained as a hollow or become infilled with new material and so form a natural cast of the original bones. These are less than ideal, obviously, as while the general shape of the bones is still there, fine details of the surface can be hard to interpret and none of the internal structure of the bone is preserved.

Perfect, but flat

A second major type of preservation occurs when the burying substrate is extremely fine grained and also settles out with very little disturbance. For that to happen, the water needs to be very still and there needs to be little or no biological activity. These situations arise from things such as volcanic ash clouds, or hypersaline water from evaporation in pools and channels. When this occurs, even the bacteria in the water may perish and so anything that was living there or fell in will decay very little or not at all.

Low levels of decay, coupled with very fine sediments, can provide absolutely exquisite preservation of skin, feathers, scales, claws, and even muscles and other soft tissues, right the way down to the cellular level. However, with such fine grains, these sediments compress enormously as they build up and become super-thin layers. A common form of this are the lithographic limestones of Bavaria, which were mined because they were useful in printing presses as the sheets of rock were so smooth and flat. As a happy by-product, the mining uncovered various exceptionally good fossil specimens such as the famous *Archaeopteryx*. These ultra-thin beds are often given the German name 'Lagerstaetten' and these have produced some of the most important recent dinosaurs, in particular those with feathers.

Unfortunately though, the bones are generally crushed flat, so while there is a huge amount of incredible detail for things such as scales or muscles, even coarse features like the shape of the skull can be hard to make out. Lagerstaetten-type deposits are also generally quite small in area, so they tend to pick up only small animals. That's good news in one respect, as this means they don't just preserve things exceptionally well, but also preserve animals that are rarely preserved otherwise (smaller bodies tend to decay faster and are more likely to be damaged when transported). However, it means we often don't get larger animals at all.

Given how large a great many dinosaurs were, this means we get a distortion no matter what the preservation, missing big species where preservation is good, but losing out on small ones everywhere else. Things like skin and feathers can be preserved in other conditions, but they are inevitably much rarer. This is the reason that we have, for example, a plethora of feathered dinosaurs from China, where there are numerous and extensive Lagerstaetten beds, but it has taken around 150 years of excavations in Canada to find only a couple of feathered animals.

Sometimes even more unusual forms of preservation can occur, too. There is a truly exceptional small theropod from Italy called *Scipionyx*, which not only preserves some exceptional features of the skeleton, and has the bones of its previous meals preserved inside it, but also has some remarkable details of the soft tissues preserved. Most incredibly, part of the digestive tract can be seen in the specimen. The stomach and intestines are preserved as a kind of impression, because what seems to have happened is that the dead animal, washing around in water, ended up inhaling a large amount of sediment. The guts themselves may have rotted away, but there was enough material in them to have left an impression of where they once were, and the classic zigzag of intestines folded up in the lower body is quite clearly preserved.

The complete crushing of Lagerstaetten fossils is rare in more traditional sediments, but that doesn't mean that the bones are perfect. Individual bones (or whole units like skulls, or a pelvis) can suffer three distinct types of distortion individually or in combination. They can be crushed (basic compression), but they can also be sheared (one side pulled forward and the other pushed back) and suffer torsion (rotation around an axis), so that the bones are no longer the shape they should be, details such as the position of joints don't relate to one another as they should, and bones may also be broken or even shattered.

It is possible to try and work out what these things should look like and there are some exciting studies on retro-deformation – scanning fossils and then using computer programs to take things back to their original shape. These are early days for such techniques, but especially for complex and thin structures like skulls or complex vertebrae of sauropods this has huge potential for making the original forms of bones visible and, importantly, comparable.

Leave nothing but footprints

Another area where distortion is a huge issue is that of trackways and footprints. You can walk along almost any beach or along a muddy pathway and see that the differing degrees of saturation will massively change the depth and quality of your prints. Very wet and very dry substrates will give you poor resolution, while something intermediate can give exquisite preservation of every detail of your foot (or shoe). Grain size will also make a huge difference; coarse grains can give you only so much resolution, rather like the pixel count on a camera, but fine-grained sediments have the potential to capture more details.

However well they start off, tracks can also be distorted and changed by the way that they dry out or become waterlogged, before they finally become fixed and preserved. A much bigger problem, though, is the way that they can change as they are made.

Tread into some really deep and sloppy mud and your whole foot will sink below the surface and you may even struggle to pull it back out. At the very bottom, you may have left a decent footprint, but as your foot rises the mud will come back into the space occupied by your foot (and leg) and is likely to collapse and slosh around. You may end up leaving something that looks rather foot-like on the surface or at the bottom, but that is not a great representation of your actual anatomy or footwear.

Similarly, as you tread into soft ground, the force you deliver will be spread out. You might leave a near perfect footprint on the surface, but you will have disturbed the soil underneath and out to the sides, too. Tread hard enough (and remember some dinosaurs weighed over 50 tons) and that force can go quite a long way down and out. If there are layers of soil, this can lead to the phenomenon of underprinting, where multiple extra tracks are laid down below the original. Each is broader and shallower than the last as the force of the animal is dissipated, but these can preserve very well and provide a grossly misleading impression of the size and dimensions of the animal that originally left them. Occasionally there are excited announcements in the press of tracks from animals that have individual feet nearly two metres across, and these are inevitably undertracks from a much smaller, if still substantial, sauropod.

In short, footprints can be wonderfully preserved, but they can also be horribly distorted and misleading. Tracks can look much bigger (underprinting) or smaller (collapse) than they really were and therefore won't necessarily match the feet that left them, and so can be hard to interpret. They can still be extremely informative though; for example, sauropod tracks and those of ankylosaurs are pretty easy to tell apart from other dinosaurs. Even when they are distorted and in poor shape, it's fairly easy to be able to identify them and be confident that those animals were around, and so be able to look at the spaces between tracks to get data on how fast they were moving, or see if there was a single animal or multiple individuals, perhaps in a group.

Happily, palaeontologists are also working on various forms of retro-deformation for tracks. In this case, the original data is often essentially lost, but by understanding how and why different types of tracks collapse under different conditions, it is possible to work out what changes might have happened to a track and thus what it may have looked like as it was generated. Slicing up some fossil tracks that have plunged through multiple layers and then digitally putting them back together can show the way in which the various bits of the substrate must have been pushed around, and give an idea of how soft and pliable things were when the animal first walked through it.

This can then be replicated experimentally with animals walking on surfaces of varying consistency, and birds from guinea fowl up to ostriches have been used for this. Obviously these best replicate theropod dinosaurs, but are an excellent proxy at least at smaller scales, and have helped us to understand what is happening as animals get to grips with soft substrates.

Combined with this, we can now take X-ray videos of animals as they walk around and trace the exact position and angles of their toes and feet as they go into and come out of the mud. Understanding tracks as structures that are produced in 3D and dynamically adds a whole new level to our concepts of them and ability to interpret them. Finally, we can replace mud or sand with very small seeds or similar particles, and use computers to track them as they move in response to our experimental birds.

While a little coarse compared to sand grains, such particles serve as a good substitute, and the ability to trace thousands of individual ones moving in three dimensions in response to the foot gives unparalleled insight into track production and how we will be able to interpret them in the future. This is a rapidly developing field and expect much more from it soon; we may yet be able to restore some incredibly distorted tracks to their original conditions, and then be able to link them to specific clades and work out with much greater accuracy which species were leaving which tracks and how they moved.

Footprints are one branch of the field of ichnology – that is, any kind of trace fossil. Traces are remains of what animals did, rather than body fossils, which, as you may imagine, are what organisms once were. So traces include things like resting marks where animals sat down; nests or tunnels or other marks of digging; bite marks on bones or shells (or even on plants); and some biological leftovers such as fossilised vomit, faeces and eggs. The two can occasionally intersect, too – there is a skeleton of a small horned dinosaur preserved with one of its footprints, for example, and we have various fossils of small theropods on nests of eggs (some of which contain embryos), so there are both multiple body fossils and trace fossils there.

Ichnology has been something of a neglected field (and worse, some gross over-interpretations have been made off very limited fossils), but it has been rejuvenated in recent years. The integration of studies like those described above for tracks, with work on the mechanics of dinosaurs as living animals and an increased understanding of the biology of living animals, allows us to glean much more from things such as eggshells and bite marks.

Even so, there are still huge gaps in our knowledge of taphonomy. We have an increasing idea of how and why certain things preserve the way they do, as there is a developing field of experimental taphonomy where bodies of animals are allowed to rot in various conditions and monitored, so that we can see the different effects of an animal drying out before being buried in water, versus one that was underwater and then dried out. There is particular interest in soft tissues and how these break down, or more accurately, under what conditions they tend to survive.

Recently there have been dramatic (and not uncontroversial) claims of original soft tissues and biomolecules surviving from the Mesozoic. Traces of cartilage proteins have been recovered from, among other things, *Tyrannosaurus*, which are apparently very (very) late surviving tissues. They have undergone a fair bit of decay and there is nothing beyond fragments, let alone whole strands of DNA or intact chromosomes. Even so, the possibility of such things surviving this long over a few million years seemed to be extraordinarily unlikely, so to have them twenty times older than this is potentially very exciting.

However, it would be remiss of me not to point out that there have been some strong criticisms of these claims, and the obvious risks of contamination are there. Dinosaur fossils are after all in the field and living animals are living and dying around them, and bits of their anatomy and decaying tissues could potentially get into and around the bones of fossils. Either way, though, this has generated a massive amount of interest from taphonomists, who are seeking to understand how these things might be preserved for this amount of time and also to help target where we might search for more and perhaps even better preserved soft tissues and biomolecules.

Dinosaurs by the dozen

When trying to work out the process that might have led to the burial of dead dinosaurs, we can literally wash the bodies of animals downstream, or look at how real-life incidents such as flash floods or animal migrations across rivers leave patterns of bodies, and determine what generally happens to them. Seeing what patterns are consistent, or which turn up only under certain conditions, again effectively allows us to rewind the clock on fossils as they are now and to work out what was likely going on in the past. Tracing the taphonomic history of a specimen can be important in order to work out under what circumstances it came to be preserved, and by extension what this might mean for it as a living animal.

For example, we know that various types of sand dunes and sand deposits are laid down by winds and that this can be very quick. So if several skeletons are preserved together in these deposits, then they were probably together when they died. Wind, unlike water, is not normally capable of transporting a whole animal, though might push them a little closer together, but a river could easily accumulate multiple bodies from miles apart and wash them up on a single sand-bank, giving the illusion of a group that died together in one spot.

Major events that can kill (and in particular, bury) are called 'mass mortalities', and are an area that we understand pretty well. Palaeontologists long ago found sites that yielded huge numbers of fossils, either a mixed bag or pretty much all from one species, and these are amazing finds either way. Getting whole faunas at a single site can save a lot of searching and getting entire populations of a single species is every palaeoecologist's dream, so they can provide huge amounts of information at once. There are, however, also biases here – some individuals or species are much more vulnerable to die in a drought or a flood, depending on their respective abilities to survive without or in water, so you can't take these at face value all the time.

Some can be confusing, too – I have worked on a huge site in China, one of the biggest in the world with a single quarry that is some 300 by 30 metres, and contains thousands of individual bones of a truly gigantic ornithischian called *Shantungosaurus*. The bones are all disarticulated and jumbled up, but with only limited sorting of big and small bones, which generally implies that these were dragged up and moved in some fast-moving water before being dumped. The fact that they were all separated, however, implies that these bones had been sat out on the ground for some time, since even the most violent of rivers and flash floods can't dismember carcasses at every joint on each of dozens (and perhaps over a hundred) animals.

However, if they were all sat out lying around as a result of some-thing like a drought or even a drowning event, sufficent for them all to decay, we would expect some large predators to have taken the opportunity to feed on them, or lots of other animals to have died at the same time. Unusually though, there's no evidence of either of these, as there are no bite marks on the bones or shed teeth from large contemporaneous carnivores, or evidence of other species, aside from a handful of bits among several thousand *Shantungosaurus* elements.

In short, it is hard to say quite what happened here, but clearly some-thing killed a huge number of animals, each of which may have been 20 tons or more. Further work on mass deaths in living groups (or recently extinct ones where changes over time have been less dramatic to conceal the original events) will hopefully help resolve this question among others. At least some dinosaurs (most notably a number of horned dinosaurs) are known from multiple bonebeds with huge numbers of individuals, suggesting both that they often lived and/or travelled together, and that they may not have been the best swimmers.

One problem with bonebeds is that if things are disarticulated, it can be hard to put things back together. Even if there are only members of a single species present, knowing which skull goes with which arm or set of vertebrae is almost impossible, so while we may have an excellent idea of the anatomy of a species, we may not have a single complete individual.

Worse, if there are multiple species present, we may end up putting bones together from two different animals without realising it. These are called chimeras after the mythical Greek monster with the head of both a lion and a goat, an eagle's claws, and a snake for a tail. Hopefully palaeontologists haven't done anything quite that dramatic in error, but we do know of cases where two or more species have been mixed up on the assumption that all the bones from a site belonged to only one.

These generally get resolved, as we have ever better understandings of the anatomy of all kinds of dinosaurs and so it becomes apparent that a limb is probably from one kind of theropod when the neck is from another, or a more complete specimen turns up that shows the chimera to be in error. However, it remains a disquieting thought that we may have a handful of chimeras knocking around in the scientific literature, or even on display in museums, without realising.

Safe and sound in the museum

Turning to museums, these are perhaps not the perfect bastions of fossil collections that we would like them to be.* Certainly disasters can befall them and various key fossils have been lost to various air

---

* Incidentally, we have absolutely no idea how many dinosaurs there are in museums around the world, even complete ones. Most museums don't even have digital catalogues of specimens and no one keeps a running total. We don't even know how many there are of things like *T. rex* even though it's a hypercharismatic animal and there are not many of them.

raids from both sides during the Second World War.* Most famously the original material of *Spinosaurus* was held in Munich until it was destroyed by allied bombers, and the small early sauropodomorph *Thecodontosaurus* from Bristol was taken out by an Axis raid. In both cases, new fossils have subsequently been discovered and we still have the original records and descriptions of them, but they remain major losses. Who knows what information they had that is now gone forever and cannot return.

Specimens get lost in museums, believe it or not (more rarely whole dinosaurs, admittedly), things can be accidently dropped or broken, specimens get stolen, and even fossil bones can decay. Various things have slowly disintegrated over time and it's not as rare as we would like that curators open drawers to discover that what was once a fossil bone is now little more than a pile of dust.

Most famously, the great American palaeontologist E.D. Cope once described a truly colossal single sauropod vertebra (actually not even that, only part of one) and named it as a new genus *Amphicoelias*. For a long time, it was considered likely to have been the largest ever known sauropod, though Cope's several descriptions of it were all short and somewhat vague, and even self-contradictory when it came to a few of the measurements. At some point during (or perhaps even after) its journey into a museum collection, it was either lost, or perhaps more likely given the fragile nature of large sauropod verte-brae, it fell apart. Either way it is lost to science and to this day it is a source of mystery and frustration that we don't know more about it and are unlikely to ever know more.

Even at their best, though, both museums and researchers can make decisions that seem good at the time and terrible in hindsight. It used to be very common practice to mount skeletons with huge iron frameworks that were physically bolted onto the fossil bones. Holes would be drilled through them and metal welded into place to hold them up. They looked great, and as far as researchers were concerned, critical information on the anatomy of the bones was preserved. Now,

---

* Accidents do happen, of course, to individual fossils and disasters can affect whole museums, such as a fire in Rio and an earthquake in Japan, both of which damaged collections that included dinosaur fossils.

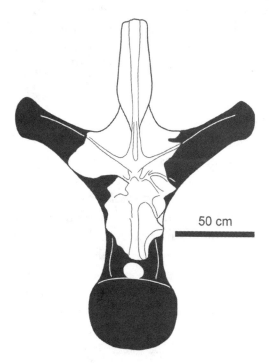

Line drawing of the giant, and lost, single *Amphicoelias fragillimus* bone. The white area represents the original material and the shaded parts are reconstructed. Quite what group this belonged to is uncertain but it was clearly a giant animal. Illustration by Scott Hartman based on the work on Ken Carpente.

of course, we know that we can get details of growth and physiology of these animals from the preserved cellular structure of the bones, but much of that has already been drilled away, or the practice has weakened the bones making them more fragile rather than more stable.

Similarly, some glues used to hold bones together turn out to have some very odd properties after decades or even centuries, and various adhesives have caused considerable consternation when they no longer hold fossils together, or become opaque and cover up the very features they were supposed to be preserving.

Other things are somewhat harder to justify but have still happened.

The legendary Berlin specimen of *Archaeopteryx** is known for its incredible preservation of both a whole skeleton and a wonderful spread of wing feathers. What is less well known is that for the first century it also sported an incredible set of leg feathers as well. However, anxious to get a better look at the bones of the leg, these were deliberately destroyed to reveal the bones. There are some poor photographs and old casts that remain from before this was done, but otherwise that remarkable record of the feathers was wilfully destroyed. It seems incredible but is sadly true, and leads to the obvious question of what other appalling decisions might have been made at some time leading to the alteration or destruction of fossil data that we can never effectively claw back, or may never even know that we have lost.

Good record keeping (be it from taking photographs, making casts of specimens or simply writing things down) is designed to offset such possible errors and retain the maximum amount of information. Still, these can also go amiss and often things were never recorded properly. A great many important dinosaurs were collected back in the late 1800s or early 1900s, when record keeping wasn't a top priority.

Finding the bones was the key and telling others where you got them from could lead to rivals from museums or other fossil collectors and dealers finding these sites and making the most of your great discovery, so people were rather more circumspect about their origins. Vague descriptions of which side of a river they were from, or how far from the nearest town, were common, and some might not arrive with even that much.

Now that we are looking to try and reconstruct accurate and detailed evolutionary histories of dinosaurs, it's critical to know if a particular specimen was younger or older than another, but it can be very hard to track down the exact site a skull came from a century or

---

* The specimen is housed in Berlin, but like all specimens of *Archaeopteryx* it actually comes from the Solnhofen fossil beds of Bavaria. Some fossils, such as specimens of *Archaeopteryx* and *Tyrannosaurus*, end up with nicknames as this becomes a useful shorthand for scientists when talking about them, rather than trying to remember museum designations such as BMNH 2381 and RTMP 77.01.324.

more after its excavation. Happily, some intrepid palaeontologists have been working on this issue and, armed with maps, letters, notebooks, rock samples and a keen eye for the detritus that comes from a major dig, have relocated many lost quarries, allowing various animals to be put in their proper geological and evolutionary contexts. However, there are many more that have not been identified and we may never be able to find them; the job is only likely to get harder as time moves on.

In short, for all that the fossil record provides us with huge numbers of specimens, we have a frustrating lack of data in key areas. Bones come without soft tissues, or jumbled up or with pieces missing, and if there is soft tissue, the bones are usually crushed. We have huge numbers of animals from deserts and almost none from islands or rainforests. We lack small animals and very old ones, and most species are still known from only a single skeleton that can hardly help to tell us much about differences between sexes or how things grew.

There is the massive upside that every year sees us discover more specimens, find better sites, improve our ability to excavate specimens and to preserve them (in the field and in museums), which can only add to our knowledge. It remains painful, though, to consider what has been lost in collections, or lost in the field, or was simply eroded from a hillside hundreds or thousands of years ago. We may find new fossils, but there will never be any more dinosaur fossils laid down and every one not collected is gone forever.

We can still do a huge amount with what we have, however, and the range of places and fossil localities that have yielded dinosaurs do allow us to get a real idea of what they were doing for 150 million years across the entire Earth.

4

Diversity

As a PALAEONTOLOGIST, perhaps the greatest leap of faith required to watch the average dinosaur movie is the fact that they only ever feature species that are well known to researchers. Of the tens or perhaps hundreds of thousands of species of dinosaurs that existed in the Mesozoic, incredibly, only already named species (and not only that, but almost exclusively famous ones!) have survived/been recreated/have got through a time portal, or whatever.

There are about 6,000 species of mammal alive today on Earth and 14,000 reptiles, and another 10,000 are birds.* Modern land-living vertebrates are very diverse and this is a single snapshot in time-many more species will come and go every few million years-so it's fair to say that over the last 50 million years or so, hundreds of thousands of mammals, birds, and other reptile species have probably existed. This stands in rather stark contrast to the dinosaurs, who in total now have about 1,500 or so known species, and they were around for the best part of 200 million years.

For all that we have described and named huge numbers of dinosaurs, we have likely barely scratched the surface of their actual diversity through the Mesozoic. This begs the obvious question of how many dinosaurs species there really were and how many of them can we find?

---

* Though remember that since birds are dinosaurs, and as dinosaurs are reptiles, then birds are ultimately reptiles. For simplicity's sake though, here I am referring to the lizards, snakes, crocodiles, turtles and other 'traditional' living reptile groups.

Diversity and disparity

This simple comparison to living groups does mask a few complexities. Compared to the mammals, for example, dinosaurs (birds aside) had perhaps only a few members capable of true flight, none that were fully aquatic, and only a handful that were likely semi-aquatic. So while the mammals do have plenty of species, the inclusion of bats, whales, manatees, seals, otters and the like is a bit of an unfair comparison.

It's also true that birds, reptiles and mammals are dominated by small species and most are well under 50 cm long and 1 kg, whereas we have almost no dinosaurs that small. Bigger things tend to be less diverse and so we might expect on average fewer dinosaurs than mammals or birds for any given period of time. Still, that hardly balances out the phenomenal amount of time for which dinosaurs were around and must surely mean that we have very few of the true number of species identified.

Number of species is only part of the story, though. We might have huge numbers of species of a dinosaur group, but if we don't find the really odd member (the carnivore among herbivores, the burrower among climbers, the glider among walkers), then we have a very false impression of what they were like. It is remarkable that compared to many other groups, living and extinct, the dinosaurs have very few species that could dig or burrow and few that seem well suited to climbing in trees. The latter is an issue in particular, because animals that live in trees perhaps unsurprisingly are not good candidates to become buried when they die and become fossils.

While it may be unfair to compare dinosaurs to mammals or lizards, for instance, we do see huge numbers of the latter living in trees and few dinosaurs. So their absence is much more likely to be from a lack a fossils preserving them, rather than a genuine absence. Similarly, large fossils are easier to find than small ones and they tend to preserve better, so while there is a real bias towards large species of dinosaur, we are probably over-looking large numbers of small ones. However, there is a great difference between what species we know of and what they represent in terms of probable diversity (number of species) or disparity (range of form).

Imagine you have a bag of photos of mammal species and you take them out one by one at random. Pretty soon you'd have a lot of rats and mice, since rodents make up about half of all known mammals. It would also be fairly clear that they were quite similar to each other – they might be high in diversity, but they are fairly low in disparity. Finding additional rodents will tell you that there were lots of them about and give you a better picture of diversity, but even with some squirrels or a beaver, it tells you mostly that rodents were small, herbivorous and generally ran round on all fours.

On the other hand, it is likely to take you a very long time to pull out an elephant (three species), or giraffe (one to three species depending on who you ask) or an aardvark (one species) or platypus (one species). These animals are low in diversity, but adding them to your pool of discoveries hugely increases the known range of disparity of mammals. Finding more species of rodents tells you (relatively) little; finding a giraffe or elephant is a big deal.

The diversity of dinosaurs

On that topic, this is a good point to cover some of the major groups of dinosaurs and what they were like. Some groups of dinosaurs are known by far more fossils or have been studied much more than others and, as a result, a lot of what we know can be anchored on a few key groups. I don't want to burden you with endless technical names and relationships, but there are some that will come up so often they're worth a mention, although there are a fair few to wade through. Dinosaurs have been intensely studied for over 150 years now and were around for the best part of 200 million years, so it should probably not be a surprise that they evolved a wide variety of different evolutionary groups across the continents.

The theropods are the most diverse in this regard with some dozen major groups (plus the birds, of course) and even within these groups there are some important branches that are worth recognising. There is inevitably a big variation in the quality and condition of fossils in any given group and the number of species known, so it's to be expected that some get leaned on more than others for research, but this doesn't mean that they become famous.

The extraordinary diversity of dinosaur body plans. Top,
the theropods, middle the sauropodomoprhs and bottom,
the ornithischians. Although there is clearly some consistency
within groups, there is also enormous variation across these
animals and they still represent only a fraction of the named
species. The dinosaurs are not to scale with each other
within or between these groups or the smallest would be
dots next to the largest. Illustration by Scott Hartman.

One of the worst represented theropod groups, low in species and good fossils, is the spinosaurs, yet especially thanks to the *Jurassic Park* series, most people have heard of *Spinosaurus*, whereas the also eponymous *Coelophysis* of the coelophysids is not a household name but is represented by dozens of complete skeletons that died together and remains a very important genus for research.

- The coelophysids were an early branch of theropod evolution and were still small animals, only a couple of metres long (although at least one very large one is known from the Triassic of Germany). Lightly built, and slender, they had rather long necks and small heads. Early theropods like this can easily be confused with the dinosauromorphs and indeed the original quarry that yielded so many specimens of *Coelophysis* was eventually shown to have a few dinosauromorphs in there, too.

- The ceratosaurs were another early branch, but these persisted right into the Cretaceous, thus outstripping their nearest relatives for longevity. These were typically much larger than coelophysids and included several species that were over 10 metres long. *Ceratosaurus* itself is famous for having three little horns on its head and is less well known for the bony armour in its skin, a rarity in theropods. Later ceratosaurs from the southern continents called abelisaurs had especially large heads and extraordinarily small arms. They were even more reduced than those icons of small appendages, the tyrannosaurs, though the abelisaurs retained four stubby fingers. One odd offshoot of the ceratosaurs seems to have been early adopters of a vegetarian diet, quite at odds with the rest of the group.

- The allosaurs perhaps match most closely what comes to mind when considering a large theropod. Many were 6–10 metres long with sharp teeth, sharp claws on the arms and presumably feeding largely on whatever they wanted. Often overlooked is the fact that many had small bony crests over the eyes.

- The spinosaurs are especially famous thanks to the utterly giant *Spinosaurus* with its huge bony sail along the back and crocodile-like head. However, while the spinosaurs are truly remarkable, even set against other theropods, they are a fairly minor branch of a theropod clade called the megalosaurs, which generally had rather large arms and claws (though the spinosaurs also exaggerated this trait). The megalosaurs may also have in their ranks the earliest theropod with feathers, although from that point on, all theropods had at least some feathers and many were as covered, if not more so, than modern birds.

- The namesake of the compsognathids, *Compsognathus,* was the first small dinosaur to be found and is famous for being used early on to link dinosaurs to the origins of birds. However, the compsognathids as a whole are a group that is little known with very few representatives in Europe and east Asia. None of them were very large, and especially not compared to some of their more illustrious and near relatives.

- The tyrannosaurs include the most famous dinosaurs of all. While *Tyrannosaurus* itself remains an iconic figure, this was merely the last of a great clade that survived for 100 million years. The later tyrannosaurs were large and well-built, huge-headed, tiny-armed animals with a literal bone-crushing bite, although the earliest forms were fairly normal and small by theropod standards and had none of these exciting features, except that many had a large bony crest in the middle of the snout. Some later tyrannosaurs were also rather long-snouted and were quite slender and lacked the bite power of their more famous contemporaries.

- The tongue-twisting ornithomimosaurs (the bird-mimicking-reptiles) were probably the fastest of all dinosaurs and essentially looked like ostriches with tails. These long-legged sprinters had small heads perched on the end of a long neck, with long arms and especially legs. They showed a number of adaptations in their anatomy, not only for high speed but also for efficiency, and would presumably have been a difficult catch for most other theropods. Early species had teeth (indeed one, *Pelicanimimus,* had a huge number of teeth and is thought to have been a filter feeder), but most had a beak and they appear to have been one of a number of theropod lineages to have gone vegetarian.

- Another plant-eating theropod group were the therizinosaurs (which evocatively translates as the 'scythe reptiles'), so called as several species had tremendously long claws on their hands (getting on for a metre, which even on a big animal is colossal). Built like squat ornithomimosaurs, these too had small beaked heads on long necks, but with rather bulbous bodies and a bird-like reversed pubis. We still don't know what their huge claws were for (though some were more modestly endowed in this area), but perhaps they were used to ward off predators since most of these animals would hardly have been well placed to flee.

- One of the most unusual group of dinosaurs was also one of the smallest and: these were the alvarezsaurs. Like ornithomimosaurs only writ small, these animals had incredibly unusual hands. Whereas the abelisaurs and tyrannosaurs reduced their arms to little more than nubs with no clear purpose, the alvarezsaurs made their arms short, but they were very powerfully built and mounted a huge claw on a fat finger on each hand. At least one of these had only one functioning finger, the others being lost. Oddly enough, the combination of being a long-distance mover with a very powerful arm has a good analogue among modern mammals – those that specialise in eating ants and termites – and it seems that alvarezsaurs were specialised insect-eaters.

- The oviraptorosaurs are the first of the groups that were especially bird-like; these also went for beaks (if in concert with teeth in several species), and many of them seem to have been vegetarian or at least omnivorous. Most of these animals were fairly small, being turkey-sized up to something that would be chest high to a human, but a few would have been two metres or more and one incredible animal from Mongolia would have looked a large tyrannosaur in the eye. Most recently a new and extraordinary branch of theropod has been described called the scansoriopterygids.* These are known only from small and frustratingly incomplete specimens, but all

---

* Yes, that's an even more complex name than ornithomimosaurs, and no, there's not much I or anyone can do about it. There are in fact huge tomes of rules about the naming of species and the groups to which they belong, and while it can result in some names that are unforgiving on those not familiar with them, they are at least often descriptive and so useful. Provided, of course, you know enough Latin, Greek and various other languages to pull apart the name and work out how it relates to the animal at hand. Still, for every stegosaur ('roof reptile', after the plates on its back) there's an allosaur ('other reptile', so called for being different to the other dinosaurs already found).

seem to be arboreal and while one has one of the longest tails of any dinosaur, another had one of the shortest (and some very odd teeth) and a third, *Yi*, was a near flying-squirrel-bird hybrid with a wing membrane of skin and feathers.

- The troodontids and dromaeosaurs as mentioned in the previous chapter are so close to the origins of birds that we still don't know exactly how the three groups relate to each other, and by extension they are extremely similar in a number of ways. Both the troodontids and dromaeosaurs were mostly small carnivores, though some dromaeosaurs got to five metres and at least a few troodontids have been suggested to be herbivores. Both groups also produced animals that were apparently gliders that could have moved around the trees effectively. Famously, these animals had a retractable raptorial claw on each foot, which would be held clear of the ground when walking, but brought into play when attacking prey or a threat.

Finally, while I don't want to dwell on the birds of the Mesozoic that lived alongside the dinosaurs, it is worth mentioning that these were also extremely diverse. Although by the end of the Cretaceous some representatives of modern lineages (including perhaps ancient ducks) were around, there was a wide diversity of birds living alongside the dinosaurs. Early forms were very much like the troodontids and dromaeosaurs, with long bony tails, sharp teeth and clawed hands, but beaks appeared in multiple lineages, the tail shrank and the claws vanished. Some highly aquatic birds evolved, which were virtually wingless and swam with their back legs, while others were ocean-going as soaring albatross-like fliers.

And so onto the sauropodomorphs–easily the least diverse of the three major groups and also the most similar in general overall form. Although the early prosauropods were bipeds (in fact, they seem to have been born as quadrupeds and then as they grew became upright), even here the general pattern of large size, small heads on a long neck and a robust body were in evidence.

- Among the early sauropodomorphs there is little to separate them from one another in terms of their general anatomy. While there were plenty of species out there and variation in head shape and neck length, for example, they are otherwise very similar (at least compared to the sauropods that followed) and they did not last long, going extinct at the end of the Triassic. It is worth mentioning both *Plateosaurus* of Germany and *Lufengosaurus* of China, each known from huge numbers of specimens, but each also criminally understudied given the potential information available in such a collection.

- The earliest sauropods appeared in the Triassic but it was in the Jurassic that they peaked in their diversity, producing a wide range of forms, and in the Cretaceous, the largest animals ever to have walked the Earth. The first sauropods are not grouped into any major clades, but represent some branches that didn't go very far, though they included some oddities such as the Chinese *Shunosaurus*, which had a tail club like the more famous ankylosaurs. Even these relatively small animals (by sauropod standards) were huge; even the earliest sauropods were in the 6–10 metre range, although much of that was neck and tail. These were bigger animals than had ever appeared before on land.

- Sauropods may never have got more extreme in form than the Chinese mamenchisaurs. While long necks might be characteristic of the clade as a whole, these took this to a new level. Animals that were around 26 metres in length had a neck that was 13 metres – fully half the animal was neck – and there was still a considerable amount of tail, too. They also had some of the longest bones of any known vertebrate – extensions of the neck called cervical ribs could be over two metres long. Mamenchisaurs have a confusing taxonomy, as most fossils are known from a huge and mixed-up bonebed consisting of numerous animals and plenty of *Shunosaurus*, as well as various other dinosaurs. Numerous species have been named, but they should probably be reduced down to only one or two.

- The most famous sauropods are those that belong to *Diplodocus* and its relatives called the diplodocoids. There are several major branches within this large group and each has its own unique features. *Diplodocus* was the first even vaguely complete sauropod to be found; casts of the famous specimen that still sits in the Carnegie Museum in Pittsburgh were sent round the world and it is perhaps inevitable that most people's impression of sauropods is taken from this animal. The diplodocids (those nearest to *Diplodocus* itself) have an extremely long tail, even by sauropod standards, with a huge series of thin and rod–like bones at the end of the tail. The mamenchisaurs may seem to be all neck, but some of the diplodocids were mostly tail.

- The apatosaurs include both *Apatosaurus* and the now resurrected (in terms of name, if not sadly in body) *Brontosaurus*. Famously the two were initially named as separate animals, then brought together under the former name, but a recent analysis of their skeletons showed that they are distinct and worthy of two names. These animals have extremely broad necks and are generally much wider and more robust than the relatively gracile diplodocids.

- Also showing a dramatic change in the neck were the dicraeosaurs. These were something of a reversal to all the usual sauropod trends, as they were mostly small (little more than elephant-sized, tiny compared to their kin) and with short necks that were barely long enough for them to reach the ground. Not only that, but whereas sauropods generally were devoid of any interesting ornamentation compared to the various crests that turn up on theropods and ornithischians, the dicraeosaurs really break with this tradition. All of them had some form of enlargements to the vertebrae in their necks and in the case of the South American *Amargasaurs*, this was two rows of elongate spines.

- Lastly in the diplodocoids we have the rebbachisaurids, which also have generally short necks, though not as short as the dicraeosaurs. Sticking with the general theme of exaggerating one part of their anatomy, these sauropods all have extremely wide mouths and they look rather like dysfunctional vacuum cleaners. They have a broad extension to each side at the front of the jaws, and large numbers of teeth with which to process plant matter, and they seem to have been specialised to feed low along the ground.

- The famous brachiosaur's main claim to fame is being one of the tallest of the dinosaurs, though a few of the later titanosaurs were of similar stature. Standing more upright than most sauropods, brachiosaurs had very long forelimbs that give them a distinctive profile. A number of sauropods had something of a dome to the top of the skull, but none are more famous than the American *Brachiosaurus* and its near 'twin' from Tanzania, *Giraffatitan*.

- Finally, we get to a huge and diverse, but rarely heard of, group called the titanosaurs. Whereas the rest of the sauropods had their heyday in the Jurassic and were generally much reduced in the Cretaceous, the titanosaurs peaked in the Cretaceous and produced the largest of the dinosaurs. They were especially common in southern continents (with especially large species coming from the then island continent of South America). The titanosaurs are odd, because although various groups of dinosaurs (and indeed vertebrates generally) lost some digits here and there, many titanosaurs lost all their fingers and essentially walked around on the stumps of their hands. We have no idea why. One branch of the titanosaur tree, the saltasaurs, is also worth a special mention, because they evolved bony armour in their skin – something that was apparently unnecessary for their relatives over tens of millions of years suddenly became rather important to them for no clear reason.

The ornithischians are certainly far more disparate than the other two great clades, as while almost all were herbivores, there are quadrupeds, bipeds, those who shifted between the two, large and small species, and a vast variety of headgear, armour and other features on them. If you want a dinosaur that is not just huge or not a two-legged carnivore, this is the clade to come to.

- The earliest ornithischians were the little bipedal heterodontosaurs, whose name aptly means 'different types of teeth reptile'. They had one set of chewing teeth (though if for plants or both plants and small animals, it's not clear), but also a large fang on each side of the head, rather like modern water deer, though facing up not down. These animals may well have fought one another with these extra weapons, although they might also have functioned as a defence against predators given the small size of these animals at only a metre or so long.
- The thyreophoreans are the great group of armoured dinosaurs that include the spiny and plated stegosaurs and the much more bone-covered ankylosaurs. Although there was some overlap in their existence, the stegosaurs were essentially Jurassic animals and the ankylosaurs lived in the Cretaceous. *Stegosaurus* is a name and form

that is familiar to everyone, with the huge body and tiny head, not to mention the asymmetric row of plates along the back and the four tail spikes, but it is actually a very unusual stegosaur and in some ways quite different from the other members of the group. Most stegosaurs had spines instead of plates along most of the body with just a few plates over the neck and shoulders and these were in even rows. Other stegosaurs also had a colossal spine on each side of the body sticking out from the shoulders and pointing back towards the tail, something *Stegosaurus* lacked.

- The ankylosaurs essentially had armour across their whole bodies. The head was all but encased in armour plates, the neck had rings of bone in it, and much of the back, shoulders and body had a variety of embedded plates, spines and discs of bone, including big shoulder spines and a solid shield across the pelvis. Even the spaces between these had tiny bony pieces call ossicles, which gave some flexibility and acted almost as chainmail between the armour plates of a knight. One group of ankylosaurs maintained a flexible tail with usually some smaller spines along it, but others had a stiffened tail with a huge bony club at the end. Long assumed to be there to ward off predators, it's also been suggested that these animals may have fought one another with these massive weapons.

- There is also a major group called the ornithopods (the bird feet, so called because they left large, three-toed tracks) that normally moved on all fours but could rear up and use only their hindlegs. The first of these are the iguanodontids, the classic of which is *Iguanodon* with its battery of teeth and spiky thumbs. Others in this group were rather similar in general, though it is worth noting the north African *Ouranosaurus*, which has a huge *Spinosaurus*-like sail along its back.

- The hadrosaurs, more commonly known as the duck-billed dinosaurs, eventually replaced the iguanodontids. Long thought to be animals that favoured wetland areas, we now know that this was not the case and they were fully terrestrial. While these animals had beaks like iguanodontids, these were sharp, cropping devices and not flat bird-like beaks for dabbling in mud. The largest hadrosaurs were truly colossal, and were heavier than many sauropods (if lacking their huge dimensions without the necks and tails that help inflate length and height values). The hadrosaurs can be split into two main groups: those that were bare-headed and those that had large bony crests on their heads, at least some of which were connected to the nasal passages and seem to have functioned to modify and enhance their calls.

- The last two major groups of dinosaurs to cover are linked by the expansions on the backs of their heads. The first of these are the bipedal pachycephalosaurs (literally, the 'thick headed reptiles'), which had enormous domes that formed the top part of their skulls and often a ridge of spikes at the rear. These animals are rather poorly known and while we have lots of thick skull caps, we don't have much good beyond the heads, and so our understanding of their biology is limited.

- At the very end we come to the horned dinosaurs or ceratopsians – their name means the horned faces. Most famous of these is *Triceratops*, a huge quadruped with a giant frill of bone at the back of the skull and three horns, two big ones over the eyes and a smaller one on the nose. This is one side of the ceratopsians, but others went for the reverse arrangement with one large nose horn and then only small or even no horns over the eyes. Early ceratopsians had no horns at all and only a small ridge of bone at the back of the head rather than a big frill, and these were often both small and bipeds, or at least capable of shifting their walking style.

Ecology and diversity

Despite the great diversity covered here, in terms of both disparity of form and number of species, dinosaurs would seem to rank behind modern mammals and it gets worse if we include all the various extinct mammalian lineages that have appeared in the last 60 million years. Much of this likely lies in part because dinosaurs never seemed to occupy the ecological niches of very small species (more on that to come), but it's also simply because we have a very good idea of all of the mammals alive now and the fossil record for the group is generally pretty good because it is more recent.

Bearing that in mind, how are we doing in terms of the number of species of dinosaurs there are to find, and are we missing any major groups that would add to our understanding of the range of dinosaur body plans, from the tiny bipeds to armoured quadrupeds and long-necked giants, and everything in between?

In general, discoveries of any new group begin slowly, as they are new to us and it can take time to recognise them; then they grow very rapidly as we search for them and know what we are looking for, and then finally they tend to tail off. When you have found nearly all of the species, it will take more and more effort to find each new one. Plotted as a curve on a graph, you can see the rate of discovery flattening out with only the occasional bump if a revision brings back a few more species, or some previously undiscovered pocket of diversity is uncovered.

What we need then is to plot a line of the rate of dinosaur discoveries and see where we sit on the curve. If it's right at the bottom we have barely scratched the surface, anywhere on the main rise and it could go on for a long time, somewhere near the top and we may be close to having them all.

Unfortunately, various factors affect the way we collect such data and are difficult to account for. Rocks have to be available to be studied and as we uncover new fossil beds we tend to find huge clusters of new species, meaning there are lots of spikes in the data and it is hard to work out when or how often we will find new hunting grounds, or how productive they will be. The process of taxonomy itself is slow, because there are so few people working on naming species, and often it can be years before new species are recognised, or species that were named and should not have been are stricken from the record.

The sheer number of researchers is also an issue. More scientists can do more work, and so while we are finding a lot of new species at the moment, for example, this could be more to do with the fact we have more people working on the problem than that the dinosaurs are necessarily easier to find.

Many attempts have been made to estimate the total number of dinosaur species and have given some wildly variable figures. Some early estimates of the total number of species have already been superseded by the actual numbers now known, which shows that they were incorrect. Many others suggest that we are still somewhere on the curve of rapid growth, which is supported by the fact that we have been naming anything from 30–50 species a year for the last ten years, getting on for one a week.

Numbers are growing and don't show any signs of fading, and while some species should probably be struck off the list, we are finding not only new species in the field, but also previously unrecognised diversity in museum drawers and collections. It is very hard to say what this may mean for the total number of dinosaur species there are to be found, but at least we can be fairly confident that we are not going to stop any time soon and the total is likely to be way higher than the current 1500-ish species that we have identified.

One further issue that it is important to understand is the different way in which species are recognised. The one most people would know of is called the biological species concept and runs along the lines of 'a species is something that can produce fertile offspring with other members', which is used in scientific circles but clearly has some issues. Asexual species don't fit with this definition and, of course, we can't apply it to fossils at all.

In fact, there are many ways of defining species and they may be separated by their genetics, behaviour or their overall anatomy. Commonly, species can be defined by all of these – those that are different in anatomy also have different genes and behaviours, and don't interbreed, so they line up nicely. However, there are lots of species that essentially look identical but are genetically distinct, or breed at different times of year and so don't naturally interbreed, even though they potentially could.

Inevitably for dinosaurs, we must look at their anatomy alone, but that means we will never be able to find such differences. We might have the bones of what appears to be all one species in terms of their anatomy, but the biological reality might have been of multiple different animals that would be separate species if we could only get to their genes, or observe them as living animals.

More than species, we might be missing entire lineages. We have not found a truly new dinosaur group in quite some time, which suggests that while there are more species to add, we might have most of the clades. Sure, plenty of new groups get recognised, but generally because we already have a few species named and researchers conclude that they are close enough to each other and different to others to warrant recognition – these are not a bolt from the blue.

The nearest we came to that was the discovery of the bizarre scansoriopterygids (the odd little flying squirrel-cum-bird theropods), but these now appear to be an offshoot of the oviraptorosaurs rather than something totally new. The tiny alvarezsaurs were only recognised in the late 1990s, but then once we had worked out what they were, it became clear that we had found fossils of them around a century earlier, but had not realised it as these were isolated bones. Of course, we may be missing some incredible and undreamed-of species – there are strange and rare animals like the platypus or aardvark that are known from only one species and a limited area – and a dinosaurian equivalent might be very hard to find (though, as it happens, for both these examples, there are decent fossil records).

Island isolation

One area that we are certainly missing and that would be of great importance are islands, especially volcanic ones. Isolated places are where species can evolve free of some of the normal pressures of crowded ecosystems and are able to occupy unusual ecological roles, thus taking on odd appearances. They can also survive long after their relatives have died elsewhere (lemurs on Madagascar, marsupials in Australia) and also radiate into many new species (Galapagos finches, Hawaiian birds).

So islands can be incredible sources of diversity and we do see some of this in the dinosaurs. South America was an island in the Cretaceous and it housed many of the abelisaurs and some of the largest titanosaurs, for example, and what is now Transylvania was then a series of small islands that played host to dwarf sauropods and unusual bird-like theropods with two sickle claws rather than one on each foot.

However, these are but a taste of what probably once existed. The Earth is home to many hundreds, if not many thousands, of islands capable of supporting unique species at any one time. Volcanic islands like the Galapagos and Hawaii with their rugged terrain and isolated nature are hotspots of diversity. They also often do not last long – they can arise, be occupied by species, have those lineages diversify, and then sink again in a few million years. The continents have been around for hundreds of millions of years, but not so these islands.

They are also terrible places for fossils to form. There is little deposition taking place for bones to be trapped and buried at the best of times and if the island sinks, it's likely to be ultimately destroyed (after all, it is by definition in a highly geologically active area). These are locations where we might expect truly unusual groups to evolve and diversify, and to have been high and with a fast turnover, too. Their absence from the fossil record is likely to be keenly felt and yet they remain a complete unknown.

Similarly, modern rainforests are areas of very high diversity, but are also places where fossils rarely form. Here the issue is the rate of decay – the high humidity and temperatures and the general abundance of life means that even bones break down quickly. So while we would expect there to be large numbers of species from such locations, fossils tend generally to be pretty sparse, so there's another major area of loss for us.

In short, while we do have a huge number of dinosaur specimens (tens of thousands, though many are very incomplete) and these represent perhaps a thousand species from a few dozen major groups, we really don't know that much about dinosaur diversity. Yes, we likely have all the major groups that occupied the major continents, but the loss of island faunas, the limited remains from tropical forests and mountains, and the major gaps of time that we have for so many places, mean that we must be missing some truly fascinating and bizarre animals. These are areas that are extremely hard to explore, and however much research effort we might pour into prospective regions, we simply can't recover fossils that never formed or have already been destroyed.

Happily though, ongoing excavations and ongoing revisions of specimens already collected keep on adding to our knowledge. It might be very hard to estimate the total number of species of dinosaurs waiting to be found, and impossible to get a good handle on how many actually existed, but the stream of discoveries shows no signs of slowing down. Some species or entire lineages would have been naturally rare, or lived in places that left few descendants and so will be hard to find, but continued excavations are only going to increase our chances of finding and identifying them.

There is, then, a dramatic contrast between the certainty that there are numerous species that lived but that we will never find as the fossils simply don't exist, and the fact that there are huge numbers of species still awaiting discovery but we simply don't know how many, or when the flood will begin to slow to a trickle. What we do know is that most probably it will not be for many years and possibly not for many decades to come.

5

Evolutionary Patterns

THE EARTH IS ever changing – rivers move course, mountains rise and erode, islands pop up and floodplains fill with silt. On larger scales, entire continents move around and oceans and seas come and go. Dramatic changes can and do happen very quickly at times, so it is always difficult to grasp the possibilities that come with talking about 150 million years of history.

Changes to environments will have profound effects on the species and lineages living there. A rise in sea level can split populations and leave them with only a few representatives, with perhaps rather different conditions to their previous range. Alternatively, a fall in levels might mean two areas are now connected, allowing new competitors, predators or diseases to move from one to the other and dramatically altering the evolutionary pressures on local species. Such events will inevitably lead to evolutionary changes to the species affected and also to modifications to their anatomy and behaviour; they may evolve to produce entirely new lineages or, alternatively, go extinct.

The fossil record is, generally, good for looking at these kinds of evolutionary patterns, as we can examine what happens to lineages over tens of millions of years and see if they were getting bigger or smaller, or faster, or more diverse, and so on. However, unlike fossil mammals, which lived on land but are recent and so have a good fossil record, or ammonites which are old, but lived in the sea so have a good fossil record, dinosaurs are old and terrestrial. So they remain rather sparse in the grand scheme of things and that can make it hard to work out what kind of evolutionary trends were going on, and especially why.

The speed of evolution

Although we tend to think of evolutionary change occurring over hundreds or thousands of generations and taking millions of years, this is not always the case. Anything that places a huge strain on a population is going to cause more rapid evolution – as competition hots up for more limited resources with only a few surviving, it is natural that whoever makes it over the line is likely to be different to the population as a whole. The genes that allowed them to be faster, taller, more green, or whatever, will now be concentrated and spread faster; similarly, any event that splits off a population, such as a new loop in a river, can isolate a small group where selection will also act faster.

Some traits can spread very quickly or increase rapidly, such as those that give a major new advantage when it comes to breeding – the first animals to get even inklings of horns are likely to be dramatically more successful when fighting for mates, and it would be only a handful of generations before every male had horns and they were getting bigger. Combine huge geological periods with a large amount of space and varied habitats, and a biological process that can run quickly, and it is no surprise that we have the enormous variation in dinosaurs that we see.

We do know that there were some major and repeated changes to the dinosaurs that often extended across many tens of millions of years. Perhaps nothing, though, is as dramatic as the changes in size that they underwent to produce the absolute giants of the Mesozoic.

Giants walk the Earth

The dinosauromorphs and their nearest relatives were relatively small animals; only a metre or two in length and weighing a few kilos, the ancestors of dinosaurs and their earliest representatives were tiny compared to those that came later. However, they were pretty big when compared to all other animals, as only a tiny fraction of animals exceed even a few grams (think of the vast diversity of insects, for starters) and even most terrestrial vertebrates are under a kilo. We are fairly big creatures ourselves and so we operate at a scale where small animals are of little consequence, but we are well into the top 0.01 per cent of animals by size. At one level then, all dinosaurs were larger than normal, but the leap up from these ancestral forms to those that were in the tens of tons is obviously still a vast increase.

The very smallest dinosaurs were around a kilo at adult size, which compared to major groups now is incredible, and the majority of dinosaur species that we know of were 500 kg or more. It's even more dramatic when you exclude the bird-like theropods, which made up almost all of the dinosaurs under 10 kg. With the exception of a few of the earliest species, the sauropodomorphs were almost never under a ton or so in size, and only a handful of ornithischians were down in the tens of kilos, with the rest being well over. Even the bipedal theropods produced plenty of big animals and there are dozens of species known that were a ton or more.

Numerous lineages across all three major dinosaur groups independently grew to exceed a ton, which stands in stark contrast to other terrestrial vertebrates. Allowing for both living and extinct mammals, most are small and only a few land mammals get to this kind of weight (various groups of elephants, rhinos, giraffe, some of the bigger antelope, giant sloths and armadillos, and a handful of others) and the very largest came in at around 25–30 tons, perhaps half or less that of the biggest sauropods.

Dinosaurs did this repeatedly, with various groups producing large species again and again. As noted in the chapter on extinction, small species tend to survive extinctions of all kinds (minor and major) and tend to evolve faster with shorter generation times, so as a result, new lineages usually start off small. However, it is still more than a little odd that their size increased so often and so much in so many groups, and second, that, given the potential for small size in so many vertebrates (species down to a few grams), the dinosaurs never shrank down to smaller sizes.

There are, on average, some major advantages to being large and so species may tend to get bigger over time, as larger individuals will do well and natural selection will therefore favour them. Being big means you should typically do well whenever you have to compete with others to get things. If you are the largest male around, you are likely to be able to get more of the best females; if you are the largest female, you can get hold of the best-quality male and/or produce more offspring.

Some of the largest and smallest dinosaurs known. The giant titanosaurian sauropod *Patagotitan* (main) and inset, the alvarezsaur theropod *Shuvuuia* is among the smallest adult dinosaurs. For completeness, the smallest living dinosaurs – bee hummingbirds of the genus *Mellisuga* are also included. Illustration by Scott Hartman.

If there is a drought, it will be the biggest animals who are going to be able to bully their way to drinking first, and bigger herbivores will be able to reach or digest foods that smaller ones will not; while bigger carnivores can target prey outside the range that smaller individuals can. As a result, we might expect that there would be a slow and general progression towards larger and larger sizes, and we commonly see this, with group after group of dinosaurs getting bigger.

It is also remarkable that many of the largest dinosaurs came at the very end of their reign on Earth. The largest theropod *Tyrannosaurus* came in the very Late Cretaceous, and the other theropods that were nearly this size (*Spinosaurus* and *Giganotosaurus*) were also from the Late Cretaceous, while various other large theropods from different clades (the huge oviraptorosaur *Gigantoraptor* and the bizarre and giant* ornithomimosaur *Deinocheirus*) came towards the end, too.

Although the sauropods had their heyday in the Late Jurassic, the largest animals all came in the Late Cretaceous, with *Argentinosaurus* and several other titanosaurs leading the way. With the ornithischians, the largest of these was the hadrosaur *Shantungosaurus* (again Late Cretaceous), with the biggest members of the ankylosaurs (*Ankylosaurus*) and ceratopsians (*Triceratops*) being contemporaries of *Tyrannosaurus* in the very last days of the dinosaurs.

Many of these animals were also discovered relatively early on in the days of palaeontology, pointing in part to the bias for finding larger things, but also suggesting that we may have already found most of the largest species, or at least, if there are more to come, they are unlikely to exceed these records greatly.

Some animals might have been approaching the limits of what may be ecologically plausible. The biggest tyrannosaurs are estimated to be over 7 tons in weight and there are no comparably large terrestrial predators to which we can compare them. Would a 10- or 12-ton animal biped even function biomechanically? And if so, could they

---

* As a side note, lineages and species as a whole can show giantism – the phenomenon of being very large – but individuals may show gigantism – which is being an unusually large member of a species. The two are often confused in the scientific literature on size and, as a *mea culpa*, I've made this mistake too in my own research on giantism.

move around enough to catch prey, or find carcasses on which to scavenge in order to keep them going? Some analysts suggest not, but it is very difficult to know and intriguing to imagine.

Given the trends towards ever larger sizes culminating with the aforementioned giants of the Late Cretaceous, what may have happened if the Cretaceous–Paleogene (K–Pg) extinction had not occurred? We can but speculate, but there could potentially have been 100-ton herbivores and 10-ton carnivores – animals that would exceed even most whale species for size.

The obvious question at this point is: if there were such advantages to large size and continuing trends to get bigger, why were all dinosaurs not big?

If you are larger, then you need more resources and so will need to eat and drink more, and it becomes more energetically expensive to move around in the first place. There will be things you can't easily access as a bigger animal (like getting down a burrow, or climbing thin branches, or moving through dense forests), and as long as there is an ecological space that can be occupied, some species will evolve to exploit it. So there will always be smaller species.

The other major factor is that bigger animals are more vulnerable to extinction – they have smaller populations, lower diversity and generally breed slowly. As a result, there's a general pattern of a drift or drive towards large size, followed by an extinction that tends to take out the largest, and then a diversification of the remaining smaller species.

The question of ecological space perhaps also explains why dinosaurs never got much smaller than a few kilos. The Mesozoic had its full share of mammals, lizards and snakes, crocodiles, tortoises, amphibians, and their various ancestors and relatives, including numerous groups not around today. Many of these were generally small and occupied similar ecological niches that they do today.

Species can outcompete one another for resources and space, and as environments change and evolution changes species, things can move around the ecological landscape. But it is notable that there were small rodent-like mammals then as now, climbing and burrowing lizards, tortoises grazing, and plenty of others, and it seems at least plausible that while the dinosaurs occupied a huge number of niches at larger sizes, they were never able to compete successfully at the smaller scales and so never produced very small species. This is to date an unexplored area, though, and something that can be tested with good datasets of species size distributions, so is an area that may yet be resolved sooner rather than later.

Cut down to size

One area where dinosaurs generally did get smaller is on the evolutionary pathway to the birds. Certainly the nearest relatives to them (troodontids, dromaeosaurs and alvarezsaurs) were typically small and between them produced almost all of the smallest known dinosaur species. Other relatively bird-like groups such as oviraptorosaurs were also mostly small by dinosaur standards and while there were some exceptions, the later feathered theropods were on the whole much smaller than all but the earliest members of various theropod groups. Even while the tyrannosaurs, ceratopsians and others were producing a series of very large species, these bird-like dinosaurs were generally getting smaller.

The evidence suggests that there was a strong evolutionary drive towards smaller sizes in the lineage to birds and that these various groups got smaller over time. Very close to the development of true flight, this may in part have been selection to lower weight and keep the gliders in the air for longer, but that can hardly explain what was going on for the earlier animals. Smaller sizes would have sped up their evolution and clearly there were small niches that could be exploited, though why other smaller theropods never shrunk to explore them is not known.

The way that the lineage to birds got smaller is also interesting, as it features an unusual pattern of development in animals called paedomorphosis, where essentially the adults of a lineage retain the traits of the juveniles of their evolutionary ancestral species. In the case of the theropods on the way to birds, and the early birds themselves, they retain characteristics not only of young animals but even of embryos. There appear to have been multiple rounds of this during bird evolution, with developmental pathways evolving such that although the animals grow up to becomes adults, they retained proportionally juvenile features, in particular large eyes and large brains with rather short faces.

Naturally they have changed in other ways – birds are not simply flying embryos – but it would be (mostly) fair to say that birds are very highly evolved baby theropods, especially when it comes to the shape and structure of their heads.

Long lasting

One other major pattern that is common to life as a whole is that, on average, diversity increases over time. There are major dips through extinctions minor or major, and of course early on in the history of life on Earth (or for dinosaurs), diversity was inevitably low. But diversity apparently continues to increase, with ever more lineages being added to the ranks of those that exist faster than they go extinct. Again this may be something of an issue as to how we would look at such a trend. A new group can arise and diversify and then die off leaving only a handful of descendants (or even just one), but that is all that is needed to act as a representative of a group.

Looking to mammals as an example, we have plenty of groups now known from only a tiny number of species, such as aardvarks, elephants, hyraxes, manatees and dugongs, rhinos, giraffe, tapirs, the platypus and echidnas. Most of these are fairly major branches in the mammalian tree of life, but between all of these groups listed there are perhaps only thirty species alive today (for context, that is fewer than all of the living species of canids combined). Groups as a whole may be all but extinct, but that 'all but' can carry on for tens of millions of years and may flourish again one day, and in the meantime may live alongside other new lineages and other holdovers.

With the dinosaurs we do see traces of this pattern, too. The stegosaurs, for example, flourish in the Jurassic and apparently finish with the end of the Late Jurassic, but several rather incomplete specimens are now known from the Cretaceous of China. This means that a group we had previously thought to have died out, held on for rather longer than we had imagined, and could also have made it even further.

In short, aside from the small matter of the K–Pg mass extinction, the dinosaurs looked set to continue increasing the number of lineages that arose. Of course, to a degree they did – the birds diversified alongside the non-avian dinosaurs during the Cretaceous especially, and added even more groups. They also added to the overarching point as well, since birds did creep over the line and then diversified massively, producing a vast array of lineages; so the progression of ever greater dinosaur diversity seems to be a real trend. That said, why various lineages arise and why some go remains something of a mystery.

New species and lineages often begin because they are adapting to changing conditions or exploiting new opportunities. So if trees evolve to produce nuts, they will provide a form of food that was previously unavailable and any animal that can evolve to exploit this will do well, at least in the short term. Some are easier to make a reasonable hypothesis about – spotting that the spinosaurs became semi-aquatic predators points to a vacant ecological niche at some point, which was able to be exploited by a theropod that was perhaps already foraging at the water's edge. But others are rather more puzzling – not least the repeated transitions from carnivory to herbivory in the theropods.

As discussed earlier, the earliest lineages of dinosaurs were carnivores, or perhaps omnivores. A shift to herbivory for some of these is not a big surprise when lots of groups were diversifying and niches for new herbivores were clearly available. With increased specialisation for eating plants and a focus on this, we see adaptations such as larger size (which increases gut length and makes for more efficient digestion), and this tends to lead to a switch from being bipedal to quadrupedal to handle the weight better. Balancing a heavy body is difficult on two legs and limbs can only take so much load, so as lineages increased in size, it should be little surprise that the arms ended up being load-bearing and taking some of the strain.

That rapidly led to producing creatures such as the prosauropods and then later the sauropods, and the huge array of herbivorous ornithischians. However, as those lineages appeared and diversified, they clearly occupied the landscapes as the dominant herbivores of the Mesozoic, but perhaps as many as five lineages of theropods from the Middle Jurassic onwards independently switched from animals to plants as their source of energy.

That is an odd switch, as there was now lots of competition from other herbivores and plants can be tough to handle. Fibrous plants can be hard to digest and low in energy content, and plants also have all manner of defences like thorns and toxins to avoid being eaten. If you have not adapted to deal with these, it will be very tough to make a living from them. So for relatively small herbivores to break into this apparent monopoly, and to do so repeatedly, points to some unusual opportunities.

Smaller species tend to be more adaptable and evolve faster than larger ones, and there are opportunities for smaller herbivores with food sources such as buds and flowers that are high in energy and will provide enough for a small and picky animal. Still, that this happened so many times and with so many groups making a success of it is odd, especially given the abundance of ornithischians (which included some small species too), and quite what facilitated this pattern has never been explored, though it may be very hard to do so without a much better understanding of plant evolution.

There have been various suggestions in the past that the dinosaurs and plants co-evolved in various ways. It would be most extraordinary if the sheer size and range of dinosaur species did not provide some serious impetus for the evolution of the plants that they fed upon – a 50-ton sauropod is likely to provide some strong evolutionary pressures on the foliage. The most intriguing one is that dinosaurs were intimately linked to the origins of flowers, but this now seems not to have been the case. The earliest known flowers are from water plants and, as noted, this was not an environment that attracted many dinosaurs, and those were carnivorous.

End of the line

Tracing back the ancestry of various groups can lead us to some fairly clear conclusions in some cases. For example, the earliest known tyrannosaurs are all from Asia or the UK, and their nearest relatives are likewise from Asia. This points to a Eurasian origin for this group, with tyrannosaurs popping up only much later in North America (which itself points to a crossing or multiple crossings of the Bering Strait).

Others are less clear, simply because of the lack of data or confusing information. We have already covered the problematic question of the likely location of the dinosaurs as a whole, with credible candidates on multiple continents, and the fact that there was a single landmass at the time meant moving between these places was likely to be pretty easy. Similarly, the origins of the ornithischians are less than clear. In addition to the obvious issues of quite where these animals may belong on the dinosaurian tree, and potentially lacking their earliest members in the Triassic, there's not much to go on.

Perhaps the earliest ornithischian evolutionarily, called *Pisanosaurus*, is from South America, but other early animals are from China, North America, the UK and South Africa. Clearly they got around, but it makes it hard to trace back to any common patterns that might explain where their origins lie, so we don't know where they first evolved and can't easily discover this without more fossils.

The flipside of this is that some groups do not appear where we would expect them to be. The horned dinosaurs first appear in the Jurassic with the early forms in China, and at that time, this part of Asia may have been isolated as a small continent. However, later ceratopsians are common in Asia and especially North America, but for a considerable time it seemed that they never made it into Europe, even long after China had been reconnected. Huge numbers of groups (living and extinct) have made it from one continent to the other at some stage, so this is very odd indeed. Many dinosaurs did so in the Jurassic and Cretaceous and there are no obvious limitations in the biology or ecology of the ceratopsians that should have prevented this move.

Then, in 2010, the snout of a ceratopsian turned up in Hungary. Named *Ajkaceratops*, this was the first, and so far, only record of these animals in Europe (beyond a couple of bits of questionable affinities). While this does resolve the issue of their inability to get to Europe, it does raise other questions, notably why did it take so long and why are they so rare here and yet abundant in other locations? This remains a most frustrating and cryptic problem and, currently at least, an intractable one.

Longevity

With mass extinctions it is clear that something major has happened to wipe out so many species and groups, even if the exact causes of the catastrophe are uncertain. However, the extinction of an individual species or small group can be far harder to determine. While it is true that species or whole lineages can fail to adapt to changing conditions, be they environmental or biological, it remains a poor catch-all explanation for the loss of a group.

Returning to the example of the stegosaurs, although some crept over into the Cretaceous, clearly most of them did go extinct at the end of the Jurassic, but with no obvious indication as to why. Perhaps newly evolving groups of predators put them under pressure, new plant species were taking over but were inedible or hard to reach, competition from other herbivores meant they were not getting enough food, or they could not adapt to a changing climate. All of these are possible, or it could have been some combination of them (or indeed, something else).

The fact that at least some survived into the Cretaceous begs the question of why the other stegosaurs didn't survive. Added to that puzzle, for those that did survive, if they were able to compete successfully with other dinosaurs, why did they not persist even longer into the Late Cretaceous, or themselves diversify to produce many more species? Frustratingly, there's no real way of knowing, and while potentially we could look to any correlations with the reduction in stegosaurs and the rise of another group of organisms, it would be a case of correlation rather than causation. We might find some evidence to support one or other interpretation, but it would be hard to confirm or to rule out other possibilities.

One final outstanding question is what is the average duration of a dinosaur species? Naturally we would expect a huge range of variation, but an average would be very useful to know. Extensive studies of the recent fossil records of mammals, where there are numerous excellent fossils, suggest that a decent ballpark figure for the average longevity of a mammal species is about two million years. That helps to give us an idea of how often new species evolve or go extinct, given the number around at any one time, and also gives us some numbers to work from if we want to know if a group is evolving or going extinct at a relatively rapid rate.

Sadly, with the dinosaurs most species are known from only a single specimen, and many more are known from a single rock formation. As a result, that gives us only a single point in time, or a fairly narrow window, and is not really any indication of a start and end point to give a duration of a species. We have very few species known from a wide range of horizons in the geological record, and fewer still where all the different beds are dated very accurately. As you may imagine, this leaves us with no more than a handful of species for which we have an idea of their duration, and even these are not necessarily very accurate.

We are left with a fairly large question mark over how long dinosaur species might have typically lasted, and it is impossible to say much about the durations of individual species, or how often they evolved or died out. Anything from a few hundred thousand years to maybe five million could be typical. Across whole lineages things are rather easier, as we have much more data for groups rather than single species, but even this can be profoundly truncated or extended as new specimens are found or others reassessed (as with the potential for Triassic ornithischians and Late Cretaceous stegosaurs).

Overall then, our understanding of major trends in dinosaur evolution is made up of broad brush strokes, and the nuance and detail, which can only come from considerable numbers of further finds, is lacking for most of them. Happily, however, those finds are rolling in and this is therefore an area that will eventually become much more fleshed out over the years. It is likely to take a considerable amount of time to get sufficient resolution for clades, let alone species, that we can talk about these patterns in great detail, but the inexorable discovery of new specimens makes an improvement in this area an inevitability.

6

Habitats and Environments

THE STEREOTYPE OF dinosaurs being restricted to swamps comes from the old ideas of dinosaurs as giant lizards, limited to tropical places and with the larger species unable to support their weight easily on land. This myth has been perpetuated repeatedly and it seems to be one that is most firmly lodged in the minds of most people. It's an easy one to succumb to and I can find it hard not to consider dinosaurs as being synonymous with heat and humidity.

However, as far as we can tell, dinosaurs lived pretty much in every ecosystem on Earth at any given time. Although the fossil record is scant in some key areas for various preservational biases, there are records of dinosaurs in swamps, but also on mountains, in deserts, lakes and seashores, temperate and coniferous forests, and across all manner of temperatures, rainfall, snow, winds, and other variations in both climate and weather.

Across the 180-ish million years that dinosaurs were around, there were huge changes to the Earth at various times. Sea levels rose and fell, the planet as a whole or various continents were hot or cold, and the landscapes changed dramatically as mountain ranges came and went, inland seas flooded or dried up, and deserts gave way to forests and went back again.

All of this was against a backdrop of other differences that would make a time traveller feel a little odd. The moon was a little closer to Earth (it is still moving away, in fact) making nights brighter and tides higher. The plant life changed dramatically over time, with the first flowers not appearing until the Middle Jurassic and not being common until the Cretaceous, while grasses were also late to the party and never formed the kind of grasslands that are familiar today across much of the world.

Life at different sizes

All of these various aspects are important when considering where, and by extension, how, the dinosaurs lived. The range in both maximum and minimum temperatures halfway up a mountain is very different to those at the top or bottom, but which side you are on will also greatly influence the amount of sunlight and rain, and that will affect what plants can grow, how big they will be and so what food or cover may be available to you. All of this builds up to affect the ecology and evolution of a species.

This can be more important for smaller species, which have a harder time regulating their temperature or finding a key resource such as water, and cannot move around so quickly to get to a better place. However, it does mean they tend to be more specialised, since they have to stay in one place and so will be adapted to the local conditions.

Members of big species on the other hand tend to have to move around to find food and no environment is so uniform that they can afford to specialise and survive – even the biggest forests have patches of no trees, or little water, or are dominated in one place by an unpalatable species. So larger animals, and especially predators that need to cover large areas to find prey, tend to be fairly adaptable and can be found in multiple, very different climates and conditions.

Leopards range across the rainforests of Indonesia, cold grasslands and dry forests of Asia, the deserts and mountains of the Arabian peninsula, and through to African savannahs by way of some more rainforests and deserts. Elephants too (though admittedly across three species) show a similar range of habitat and area of occupation, though they are generally not found in the colder climates that the leopards can occupy. It should be no surprise therefore that sauropods, ceratopsians and large predators like tyrannosaurs might have also ranged across great tracts of land, not only as lineages but also as single species or even individuals.

The requirements of food and varying conditions must have made some species need to be migratory in order to keep eating. Similarly, dinosaur eggs needed time to mature and hatch, and this could have forced annual trips to environments with less flooding or more heat, to make sure the next generation would survive. In short, we might expect there to be two or even more species of a small dinosaur in an area with multiple climatic conditions, but one large species might both be here and also crop up hundreds or even thousands of kilometres away.

Demonstrating this for any given species is very hard. Fossil beds are generally accumulations of anything from dozens to tens of thousands of years, and within even the smaller of these timeframes, there could easily be a drought or extensive rains that fundamentally changed the regular distribution of the species, giving palaeoclimatologists a false impression of the usual conditions. Showing that one species did have a narrow distribution and another a big one is near impossible, and tying that to local habitats may be even harder.

We do have species that appear to have extremely large ranges, but these remain contentious because of the fragmentary nature of the fossil record. *Tyrannosaurus* is known from southern Alberta through to the north of New Mexico and it probably had a still wider range, but teeth from Mexico assigned to this genus while certainly from a tyrannosaur, could be from a different, if closely related species.

Similarly, there is a huge titanosaurian sauropod called *Alamosaurus*, which ranges from Texas (as you may expect) to New Mexico and Utah. However, the various fossil beds that have produced this animal are also rather separated in time and much of the finds assigned to this genus are very fragmentary and poorly known. This is the problem with such apparently widespread species, as too rarely are there fossils that are well separated in space that are from the same time and hold enough material to make a definitive comparison.

What we find wherever we look (with the exception of those examples given above), are dinosaurs and this is in part the reason for assuming that dinosaurs occupied almost every environment going. They were after all highly diverse and near ubiquitous for the Jurassic and Cretaceous and it should be almost inevitable that they would be able to colonise and adapt to any newly emerging ecosystems.

That said, they didn't have it all their own way. In 2009, a large number of new crocodiles were described from one region of Niger, which also proved to be from a fossil bed remarkably free of dinosaurs. Ancient crocodilians were much more diverse in the Mesozoic than now, but even so here were a series of species that were well adapted to living on land and included both herbivores and carnivores. It seems likely that a new ecosystem formed and the first colonisers happened to be crocodilians and not dinosaurs, and so they diversified and took over this corner of the globe.

A couple of other places are known that seem to be dominated by pterosaurs in the Cretaceous, one of which, the Junggar Basin of western China, is one I have worked on. This was a great desert with one (or perhaps several) large lakes and oases in the centre. The pterosaurs could have easily flown in, but it must have been difficult for dinosaurs to reach or feed in this area. As a result, there are huge numbers of pterosaur fossils, but dinosaurs are rare.

A few bits of small theropod are known and while prospecting there years ago, I and my colleagues found the tooth of a large theropod and a badly eroded skeleton of what appeared to be a medium-sized sauropod as well as some lizard and crocodile bits, but otherwise the bones were exclusively pterosaurian. These examples stood out precisely because they are unusual; ordinarily in the Jurassic and Cretaceous, any terrestrial fauna would be first and foremost notable for the presence of dinosaurs. In general, they dominated terrestrial ecosystems worldwide.

Differences in climate will influence the local plant species and will also have major effects on the local faunas and, by extension, the animals will necessarily be different in their behaviour and ecology with the changes in prevailing conditions. The great wildebeest migration of the African savannahs in the Serengeti and Maasai Mara is very familiar, as it turns up in almost every wildlife documentary going. But while it is normal for those animals living around the equator to chase the rains and the resulting new grass growth, down in South Africa, they (or at least the males) are non-migratory.

In the south, the seasons are much closer to a spring–summer–autumn–winter cycle and it's not as if when the rains stop in one place they have simply moved on a bit and there is fresh grass over the horizon. Instead, when it dries up and there's no more grass or other fresh leaves, the animals basically sit still and wait for the rain to return, rather than being on a constant move to find fresh food. That's a fundamental change in how these animals live over a year, but it also influences the species around them too, with different pressures on the local plants from grazing and browsing, as well as the local herbivores in terms of competition and the predators in terms of prey availability.

We might expect such changes to influence the dinosaurs, but it's hard to make any firm predictions. These kinds of patterns in living species are partly mediated by size – small animals can't travel huge distances and don't need that much food; it's a lot of effort for big ones, but they can digest things that medium-sized animal can't. But it also depends on how much food is available and that is incredibly hard to predict. Plus our data is always limited, so what may show up as a highly seasonal environment in the fossil record may be correct, but could also be a situation that persisted for only a few decades, against tens of thousands of years of a very different time.

Returning to the current African plains, while the major picture we have is of an environment of open grasslands with few trees, in the last two centuries alone this has changed twice across large swathes of East Africa. It has gone from this classic view to one of dense, dry forests and little grass, and then back again. This will inevitably have a major effect on other plants competing for light and water and those that feed on them, so whole ecosystems of flora and fauna can turn over relatively quickly, with populations of species rising and falling in response.

Certainly some dinosaurs did appear to migrate, with there being extensive evidence of this in hadrosaurs and sauropods, based on the enamel of their teeth, surprisingly. The enamel of vertebrate teeth is remarkably resistant to wear and chemical change over time (which is in part why fossil teeth are so plentiful) and this also means they tend to preserve the original biology of the animal from the time when they were grown.

Isotopes are variations of elements where the atoms have extra neutrons. They function chemically almost exactly like the normal versions, but are heavier because of those excess neutrons. Certain biological and environmental conditions can concentrate one isotope over another, and so as a result, unusual proportions of these relative to the regular isotopes can be produced. We can detect these different isotopes in both fossil teeth and the rocks in which they are preserved, and so can see if these ratios are unusual and match to the relevant known conditions. Grind up a portion of each and scientists can detect these differences and see if the animals were living and feeding in the areas in which they died.

What we find in a number of cases is that these animals were regularly moving from location to location, with differing bands of enamel in their teeth showing certain isotopic signatures coming and going with the animals' movements. Presumably they were following the trail of food, but they could also have been returning to key areas to breed, for example, and as we have seen with the wildebeest, even if some members of a species did this, it's not necessarily the case that they all did, let alone that we can apply this to other members of these groups.

Dinosaur ecosystems

The nature of habitats is driven by both the climate and also the plants themselves. Trees give height to an environment and generate shade, which will impact negatively on some other plants but can also provide places for mosses and creepers to grow, as well as a new way of getting around if you are arboreal. If there are only low-lying plants like ferns, then it will be difficult for large animals to hide, whether they are herbivores trying to avoid predators or carnivores trying to avoid being spotted by their potential prey. So all of this is very important to understanding how dinosaurs lived in real worlds, but fossils of plants can be much rarer than skeletons (since they decay more readily than bones), so how can we reconstruct the look of an ecosystem?

Happily, we can often get a decent picture of the nature of plants in a given dinosaur locality based on fossil pollen. In life, pollen grains frequently have to survive in harsh conditions with a precious payload of plant gametes, and so they have a surprisingly hard 'shell', which makes them good candidates for fossilisation. Combined with the huge numbers in which pollen grains are typically produced by plants (as anyone with hay fever can testify), this means getting large amounts of them from most fossil beds is fairly easy.

The pollen grains are tiny, but can be reliably identified, which means we can begin to piece together what kinds of plants were present and in what numbers, revealing if an area was full of forests or plains, with tall or short plants, and so on. The major parts of plants themselves might be rarely preserved, but through the study of fossil pollen (properly called palynology), we can start to understand the kind of environments in which some dinosaurs lived.

Some are better known than others, thanks to the vagaries of preservation. The famous late Middle Jurassic Daohugou and early Late Cretaceous Yixian beds of northern China are renowned for their extensive preservation of feathered dinosaurs and numerous other small and generally rare fossils, like amphibians and early mammals. However, they also yield huge numbers of tree trunks, which help to show that these were full of massive trees, and so it is perhaps then less of a surprise that there were more birds and gliding dinosaurs here than in many other places.

Gliders in particular need to be able to get high enough off the ground to make successful launches and there would be little evolutionary pressure for gliding to evolve were it not for the efforts of having to go up and down trees all the time, which makes a short flight a shortcut. As ever, though, this is more an exception than a rule, and for most habitats we have only a very limited idea about the nature of the plants there, and especially things such as the height of trees, how much cover they provided, how dark it was, and what other plants lived there.

From here, as you may imagine, it then becomes difficult to piece together how dinosaurs may have interacted with their environment and what they may have had available to eat, or what the conditions were like at various times of year.

Dense equatorial forests can be quite dark on the ground even in the heights of summer, which lies in contrast to the polar regions with their summers without sunsets and winters without a sunrise. Various dinosaur groups would have been present to experience all these conditions, but to what extent is not really known. Especially when it comes to both the Arctic and Antarctic regions, some species doubtless migrated out in the winter to return when it was warmer, but it's likely that at least some of them would tough it out through the frozen winters. We know of modern alligators that can survive through the freezing of their ponds in winter by sitting mostly below the ice with their snouts exposed above the water, and plenty of lizards, birds and mammals hibernate through harsh winters.

Fossil tree trunks from the Cretaceous of China. Although these
are on display in a public park, they were excavated nearby
and erected (with some restoration). Photo by the author.

What strategies, or combinations thereof, the dinosaurs might have
used to do this is uncertain, but some arrangement of lowering physiology, finding a good shelter underground or in a tree trunk, and insulation from filaments or feathers of some kind, could all have helped.
Whether or not we could find any evidence or this, or be able to
recognise it as being linked to a hibernating behaviour, is up in the air.

Let there be more light

Such winters would, of course, have been cold, but the variations in
light would also have had major effects on the animals in these environments. Camouflage is important for both herbivores and carnivores

that need to fit in, and the differences between harsh daylight and that filtered through dense forests can be stark. Patterns and shades that provide excellent protection under one regime can make things stand out in another. Colour, too, will be important, as seasons change and backgrounds can alter dramatically with the fall of snow or the loss of leaves from trees, for example.

The activity of the animals themselves can be critical, too – if they are active in the middle of the day when shadows are at their deepest, then strong contrasting colours are likely to be important, but if they are nocturnal, then dark shades and possibly a single uniform colour might be best. With a lack of understanding of the environment, daylight, and activity times of a given dinosaur, this all has knock-on effects for making it hard to interpret what they may have looked like and how this would fit into where they lived.

Light levels change both daily and seasonally, and there are seasonal changes in weather and gradual changes in climate. We have enough trouble trying to deduce local climates for various fossil localities without the obvious problems that these would be changing, especially over periods of tens to hundreds of thousands of years. Of course this would have major effects on the local conditions and, by extension, the plants, and therefore also the animals. Such change is to be expected and we can look at some long-term patterns and see if the climate for a given region was warming or cooling or becoming wetter or drier over time, but this can miss short-term dramatic events, which can be equally or more important in shaping an ecosystem.

We see this today with the El Niño weather patterns. This is a phenomenon that occurs every few years, where variations in ocean temperatures in the Pacific lead to sometimes dramatic changes in local rainfall and temperatures in South America especially, but even worldwide. Occasional droughts or floods can therefore follow and might vanish in surveys looking at temperatures across ten thousand years. However, these would have major impacts on local plants and animals – if you were not adapted to deal with occasional droughts or floods or wildfires, then you would probably not last long and would soon go extinct or be forced out of the region.

On the other hand, such dramatic events and shifts are exactly the kind that are most likely to lead to the preservation of the local faunas.

Droughts will kill off most animals and mean that even scavengers are in short supply, leaving ample skeletons scattered across the landscape to preserve; or sometimes floods can capture everything and similarly prevent the normal breakdown of bodies and enhance the preservation of the fauna. But that means we as palaeontologists looking back may have a misplaced impression of the local conditions.

Evidence of floods are fairly easy to deduce in the fossil record. We see limited breakdowns of skeletons through the action of scavenging theropods or other animals, and the mixture of sediments and the alignment of bones indicate a flow of water that mixed and moved things. But if most of our bones come from such layers, the temptation is to infer that flooding was common, when the truth is it may have been relatively rare but merely that it was excellent at providing us with data. Droughts will also force some animals to leave an area, while others may be relatively drought tolerant, meaning that any fossil beds laid down after such a situation would be unrepresentative both of the usual conditions in the environment and of the types of species that generally lived there.

Beyond the basics (well forested and warm, open plains with high seasonality), there is still a lot that we don't know about dinosaur habitats and therefore what this would mean for their lifestyles. As before, given the diversity that they showed and the amount of time they were around, it would be rather odd if some of them didn't evolve to be hibernators, or shed their feathers to change colours with the seasons. It is one more area of research that will likely improve as more attention is paid to these issues. Happily, many of these gaps in our basic knowledge about local plants and climates are not too hard to work out; it merely takes time and effort.

So, while palaeontologists (and especially palyonologists) are thinly spread and very busy, it should mostly be a matter of time until this becomes an area that is much better understood.

7

Anatomy

THE STUDY OF dinosaurs is intimately linked to the study of anatomy – at a fundamental level, the study of these animals is the
study of their bones. Although palaeontologists are reconstructing the
ecology of dinosaurs, their evolutionary history, relationships and even
their behaviour, all of this is ultimately founded on studying their
bones and teeth and on occasion, some impressions of scraps of skin
or other soft tissues. Palaeontologists are and remain anatomists and it
is no coincidence that a huge number of dinosaur researchers teach
anatomy to medical or veterinary students.

The medical research field long ago moved on to more exotic
pastures, but the first principles of how humans are put together
remain critical to basic medical training, and the best and most experienced anatomists these days are often those with a palaeontological
rather than human leaning. Happily, there's enough in common
between the basic construction of humans and even dinosaurs that
makes this an easier step than might be imagined, though of course
the details are rather different.

It is tempting, then, to assume that the fundamental anatomy of
dinosaurs is an area that is essentially in hand and, for the bones and
teeth at least, requires little more than some finessing and updating as
new species are discovered. The fact that there's a whole chapter on
this subject in a book on what we don't know will suggest otherwise,
but there are some nuances and odd details to the gaps in our knowledge about dinosaur anatomy.

The importance of 'basic' research

First off is the inevitable issue that although anatomical details of dinosaurs are the basis of almost all other research, as a field, it is, sadly, not considered especially glamorous or worthwhile. Scientific funding bodies and scientific journals are much more interested in studies that look at, say, the evolution of whole groups, or solving issues of how they moved or what their ecology was, or new specimens showing off skin or representing new and odd species. All of these are entirely worthwhile pursuits of course, and I am as interested in these areas as any one of my colleagues. But it does seem to be overlooked that this kind of work is essentially impossible without the basic anatomical work on dinosaurs (and indeed living species as a comparison).

Even the most in-depth and complex evolutionary study is essentially based on large numbers of previous papers and books that worked out and described the bones of various fossils. It's not uncommon for palaeontologists to cite numerous papers and studies that are well over a century old, which would be unheard of in most other scientific fields. This is the basis of our science and yet long and detailed descriptions are often shunned because they are not very exciting.

We do, though, still have many good anatomical descriptions of dinosaurs (even if we need many more) and so even fragmentary specimens can hold a huge amount of useful information. A single piece of bone can be potentially diagnosable to the level of a species, so allowing us to say if a given group was present at a particular time or place. We can make inferences about how fast an animal was, even if it's known from very little, because we have a good understanding of the fundamentals.

Short thighs but longer legs, for example, correlate well with high speed, so only a couple of bones may be enough to suggest an animal is a fast runner. A new species of dromaeosaur might be named from a partial skull alone, but given that we have lots of complete skeletons of these animals, we can be confident that it would be lightly built, have long arms, a stiffened tail, and large claws on the hands and feet. Sure, the exact details such as the length of the legs or the curvature of the claws would be missing, and these might prove to be unusual if we ever found better specimens of the species, but the basics are likely to be correct. This only works, however, when you have lots of good skeletons, and some aspects of dinosaur anatomy are quite extraordinarily uncertain.

Take the sauropods, for example. These animals are in some ways over-represented in the fossil record since the bones of animals that could be over 50 metres in length are quite easy to find (on the flipside to be fair, they might be buried less often than other groups *because* they are so big – it's hard to bury that much animal, even in a flood), so getting hold of sauropod bones is quite straightforward. However, even though we have dozens of great skeletons of these animals and they are known from hundreds of specimens representing numerous species, there are perhaps only a dozen or so well-preserved skulls for sauropods and many of these are concentrated in only a few species.

Skulls are extremely important for understanding the biology of an animal – the skull holds the major sense organs, the brain, and the main ways an animal gathers and processes food (namely using the eyes, ears and mouth), as well as having a critical role in getting air into and out of the body. Certainly you can learn a huge amount from a headless animal, but to get into real details, and especially to compare it to near relatives, you need a skull. To be so bereft of the most important part of the body for so many species of such a major clade seems almost absurd, but here we are. We probably have more good skulls of *Tyrannosaurus* alone than of all the sauropods combined, despite their having been around for 150 million years and living on every continent.

Other thin bones like ceratopsian frills (and indeed dinosaur skulls generally) are also commonly distorted and bent or broken, making them hard to analyse and compare, though at least they are typically found in large numbers. Work is being done on these to 'retrodeform' them – taking 3D scans and images of skulls and digitally pulling them back to their original shapes. When we have large numbers of skulls, we can create an 'average' of them all as a target to aim for or reflect, and copy undistorted bits from some skulls – or left and right sides – to put things back together and slowly work towards something that better approximates to the original shape. We can even look at the ways the original rocks containing the fossils were folded and altered and use that as a guide for pulling things back.

The right (above) and left (below) sides of a skull of the
ceratopsian *Centrosaurus* showing some crushing and shearing
of the bones. Tens or hundreds of millions of years of geological
processes rarely leave fossils in the exact shape they were
when the animals were alive. Photos by Jordan Mallon.

This procedure is in its fairly early stages at the moment, but if this
works well, it will open up a huge number of specimens to be returned
to close to their original forms; allowing us to answer a lot of questions
about shapes and how they changed over time, which are currently
impossible or very hard to solve because of the gaps in the data.

Missing sauropods skulls and distorted ceratopsian frills are not the
only big problems we have. Small bones are regularly missing from
fossils, as they are light and fragile enough to drift away or be destroyed
by natural erosion, or simply by scavengers taking them apart. That
means that things like toes, ankles, and the tips of tails are almost
inevitably missing. Even skeletons described as being complete generally have a few elements missing and there is pretty much no such
thing as a truly complete skeleton.*

* The exception to this are the beautiful skeletons from Lagerstaetten
deposits and some of these truly are complete, though they are then almost
always flattened into two dimensions. Complete skeletons that are *also* 3D
are all but non-existent.

For many decades it was thought that birds could not be related to dinosaurs because all birds have a furcula (the wishbone, made up of a pair of fused clavicles that we call collarbones) and no dinosaurs do. The (incorrect) assumption was that a lost feature could never re-evolve and the absence of a furcula or clavicles proved that dinosaurs could not have given rise to the birds. It turned out that this was merely an absence-of-evidence problem. We have dozens of dinosaurs preserved with furculae, including numerous different theropods up to and including various bird-like dinosaurs and early birds like *Archaeopteryx*, and even down to some early branches such as the prosauropods. Furculae were present in most or all of the theropods and at least some animals on the sauropodomorph lineage; they simply tended to be fairly small and were easily lost.

The tips of tails are especially vulnerable, since they are not held onto the rest of the body with any complex joints in the bone, major muscle groups, or even much in the way of tendons and ligaments, and small cylindrical bones are always likely just to roll off. However, tails as a whole are fairly rare and to get most of a tail, let alone a whole one, is very rare. Tails in dinosaurs are important, as they support the main locomotory muscle groups and in bipeds they also provide a counterbalance to the rest of the body. Based on some work I have done, we know that the lengths of dinosaur tails overall, and their composition, are extremely variable both between and also within species.

Quite what evolutionary pressures led to these various changes for most groups is not clear, but early dinosaurs seemed to have a fairly simple system of longer bones at the base of the tail and shorter ones further down. Various later groups added different regions of short, long or steadily shrinking bones, presumably to alter the degree of stiffness in the tail. More short bones means more joints and more flexibility, fewer long bones means fewer joints and more stiffness, and this would have altered how much the tail could move and what muscle groups may have helped to keep the animal moving.

The tail was probably also important in the various gliding dinosaurs, and keeping the centre of mass of the animal over the centre of lift generated by the wings is critical for flight. If too much of the animal's weight is in front of or behind the main point of the lift, the animal will pitch forwards or backwards accordingly. The fact is that many gliding dinosaurs such as *Microraptor* seem to have had large fans of feathers on the tail that likely acted as an additional source of lift, as well as having an effect on steering. So an absence of tails for these dinosaurs is especially important to understand how they may have flown (or for those where this is uncertain, if they could even fly at all).

Having large chunks of dinosaurs missing is something to which we are resigned, but it can be especially problematic when there are some potentially massive implications on the line. A few years ago, the legendary giant carnivore *Spinosaurus* was reinterpreted as an especially giant animal and the first and only known theropod to have become a quadruped. This was based on some specimens coming to light with apparently small legs and the centre of mass of the reconstructed animal being forwards, forcing it down onto all fours.

Here the problem of missing data is rather acute. The new reconstruction was based on a number of fragmentary and even now lost specimens, and of animals of differing ages and of varying degrees of relatedness to *Spinosaurus*. That much in itself is hardly a problem; it is entirely normal and if we want to reconstruct dinosaurs from limited remains this is about the best practice available. The issue is that if the *Spinosaurus* really is a quadruped, we might expect that their arms have been heavily modified for walking, but they remain completely unknown apart from a couple of wrist bones. Alternatively, the arms and the neck (also missing) could be massively reduced, meaning that the centre of mass is further back and it would be a biped.

Similarly, the tail of the spinosaurs in general, and their nearest relatives the megalosaurs, is not known well at all, and though the basics of the tail are not hard to get right, critically the length is essentially unknown. So quite how well this would have worked as a counterbalance is uncertain and any estimates of the total length are going to involve quite a lot of guesswork. None of this means that this was not the largest of the theropods or that it was not a quadruped, but there is so much uncertainty around the reconstruction because of the missing bits that this is merely a credible hypothesis rather than confirmed. It remains a massive gap in our knowledge and more research is desperately needed to help resolve this.

These various absences can therefore mask lots of cryptic diversity. Evolution takes some very odd twists and turns. For example, you would not immediately guess that there is a shark species out there that is predominantly vegetarian, based on their far more numerous cousins; and animals with a highly specialised diet or ecology like cheetahs, pandas and gerenuk would not be obvious extensions of the diversity of the rest of their respective groups.

In short, dinosaurs were probably a lot more weird and varied than we recognise even now, because however strange a stegosaur may be, if we have only a few of the most common species represented and are completely missing a hypothetical one with no plates — or that dug well to grab mammals from their burrows, or had super-long legs and was a fast sprinter — then we end up with an incorrectly conservative view of them and their evolutionary history.

Such things are unusual, because by definition they are uncommon. So given all the absences and biases in the fossil record, it's predictable that great swathes of the strangest animals are missing, or more frustratingly, we have their fossils but lack the bits that would show us that they had weird and wonderful lifestyles.

Some bonebeds that capture numerous different species have all of the bones jumbled up, so that we can't easily work out what bits go with which species. There's a huge and wonderful mass burial site in the small town of Zigong, in central China, where numerous Jurassic sauropods as well as some smaller ornithischians and some large theropods are preserved. There are several huge and bony clubs preserved at the site that clearly belong to a sauropod of some kind, but with two different genera well represented — the rather apatosaur-like *Shunosaurus* and the enormously long-necked *Mamenchisaurus* — it has been far from clear for a long time if one or other or both had tail clubs.

It increasingly looks as if it belongs to *Shunosaurus*, if mostly because various species of *Mamenchisaurus* are now known from other sites and while none of them are complete, no tail clubs have shown up, and these are the sort of thing that would preserve well. Still, a better resolution of this issue would be most welcome and a more complete tail with a club on the end would help better explain how these animals wielded them.

Another Jurassic sauropod has turned up since, *Spinophorosaurus* from Niger, which rather than having a club has some stiffened verte- brate at the end of the tail with some pieces of spikes. This suggests that it was something of a tail-mounted weapon as well, but there has not yet been a major analysis of how these various weapons might have functioned.

Squidgy bits

The soft tissues of dinosaurs are even more poorly known, of course, and it's not too critical to dwell on them since this is an area that is simply too reliant on new discoveries. There's always the capacity to use new techniques and contexts to reanalyse skeletons and so on, but when the fossil record of such things as stomachs and livers is essen- tially nil, there is very little to work on. However, there are at least some areas that, while they remain unknown, might radically change our understanding of some species and are worth reflecting on a little.

First off is muscle composition. In general, vertebrates have two types of muscle, red and white, familiar to anyone who has eaten chicken or beef. Animals have both, but the proportions can vary enormously. The red muscles are better at endurance and can keep going for longer, while you can get more power from white muscle, but it tires quickly. We tend to associate white muscle with birds, but pigeons and ducks that fly long distances have dark, red muscles and chickens have white meat, partially because they have been bred for it, but mostly because taking off quickly when you are a poor flier requires a huge burst of power quickly and that will give you white muscle.

Among mammals, red is near universal, though cheetahs are loaded with white muscle, but what might various dinosaurs have had? It is not unreasonable to assume that animals such as sauropods and cerat- opsians were almost exclusively red muscle, and adaptations for high speed in groups like the ornithomimosaurs imply they would have had some, if not a lot, of white muscle. But for some of the others, it's much less obvious and the implications are important.

The tyrannosaurs have adaptations to travel very efficiently, although in the case of the largest, they could not have been especially fast simply because of their size, and they are considered to be endurance animals. But the juveniles have much longer legs proportionally and could have been fast runners. If they were loaded up on white muscle, this could have made them substantially faster and changed what kinds of prey they pursued, but if so, then at some point in their growth they would have had to switch tactics and their muscle types, which would be unusual.

As for the dromaeosaurs and other near-bird dinosaurs, these would seem to be prime candidates to have loaded on white muscle to make them even faster, but that's not necessarily the case, since there is always the trade-off between speed and endurance. They would have to watch out for larger predators that might make a meal of them, or local environmental conditions could mean they had to make long journeys every few days for water, which would require good stamina. For a semi-desert-dwelling animal like *Velociraptor*, therefore, you could make a reasonable case for either extreme and the difference in performance and ecology would be quite substantive.

Desert dwellers are also the kinds of animals that may require extensive fat deposits to make it through hard times, and those living in the coldest climates have the same requirements and might need the insulation properties of fat, too. Camel-like humps have been suggested for some desert-dwelling dinosaurs with elongate spines, such as *Ouranosaurus* and *Spinosaurus*, though the detailed anatomy of their vertebrae is quite different to those animals with fatty humps, so this appears not to be the case, but it doesn't mean they didn't have fat storage elsewhere.

In the case of the Arctic and Antarctic species that did inhabit cold climes, fat may have been essential. It's reasonable to infer that it was present, but the change in appearances could be very dramatic. Hibernating bears can lose a full 50 per cent of their mass over the winter, and while not all of that is fat burning off, the difference in appearance between a pre- and post-winter hadrosaur could have been extremely dramatic.

One last area worth touching on is that of the digestive system. We have a fair idea of what most dinosaurs were eating based on their teeth and skull mechanics, and then direct evidence from things like stomach contents and coprolites, but that still goes only so far. There is one incredible fossil from Italy, the tiny theropod dinosaur *Scipionyx*, which preserves some incredible information on the intestines, but it provides only so much information.

With *Scipionyx*, the individual was buried in mud, and it seems that a lot of this ended up in the gastric tract of the animal as a result. Although the stomach and intestinal tissues rotted away, the mud is preserved in place and so an impression of the digestive system remains. It's a truly incredible specimen that has (unlike far too many) been described in enormous detail, and it tells us that in most ways this dinosaur had a fairly unremarkable system of a stomach and then a series of intestines.

Nothing too intriguing there, then, however incredible it is that it's been preserved, but preservation of some of the much larger herbivores would be more revealing, since the complex processes of breaking down tough plant material has led to a far greater diversity of digestive systems in living animals. It's quite possible than many of these dinosaurs had various combinations of fermenting chambers or divided stomachs, and any understanding of this (assuming it was present) would greatly inform our understanding of what they could and could not eat, and how effective they may have been at extracting nutrients from them.

Functions of features

Even when we have very complete and good sets of information on various anatomical features, it can be difficult to determine quite how they functioned, or even if they had any function at all. Natural selection suggests that features which help an organism survive and reproduce will be kept and those that incur a cost and reduce the chances of survival will be lost, but there is a potential third way. Some features or details of them may have no real benefits, but also no real costs; they are minimally expensive in terms of energy required to grow and keep them, but then they don't necessarily help out either. These features may be said to be under neutral selection and in theory these can more or less come and go with no real effect on the lineage. However, with such limited data available to palaeontologists and with an eye for adaptations that can help us understand how dinosaurs lived and evolved, there is a potential risk that such structures will be assumed to have had a function and then we will struggle to explain how these might work, when the truth is that they didn't do very much at all.

Staying on the theme of the arms of various predator theropods, we come to a more fundamental question about their anatomy and function: what are they actually *for*? Endless debate has raged about the reduced arms of tyrannosaurs and how they may or may not have functioned in everything from prey capture, to grooming, to stabilisers during sex – but all of these overlook the fact that reducing arms in an animal that is basically a walking head makes a certain amount of sense, but keeping long arms with claws in many others does not.

When we reach the derived small carnivores, they tend to have very long arms that could reach the mouth, coupled with long feathers on the arms, which themselves might have had a number of different functions. These are not a great concern, but for animals such as the allosaurs, ceratosaurs and megalosaurs (and the early tyrannosaurs that don't have reduced arms) it is far less obvious. The arms are quite long, the hands quite large, there are good-sized and curved claws on them and their anatomy suggests they had a good amount of muscle on them, though with a somewhat limited range of motion.

Their overall size and the size and curvature of the claws suggests they were functional, or they would have been reduced as in so many other lineages, but that function is far from clear. They are too short to reach the mouth and being slung under the body would not have been much use to hold down prey (alive or dead) and then try to bite it. If charging after prey, you want to strike at it as soon as you are in range, but that would involve using your head, not trying to overtake it and then grabbing it underneath you with arms that have a limited range. Other lineages such as the large tyrannosaurs and the abelisaurs apparently did fine as carnivores with tiny arms, so clearly these were not essential in prey capture, or survival generally. This remains something of an anatomical mystery and one I have been hoping to explore, but have never found the time to engage with fully (a common refrain for many an academic).

In summary then, while the foundational anatomical work on dinosaurs is extraordinarily strong and probably the single best understood area of their biology, there remain some large gaps in our knowledge. Still, with the exceptions of the soft tissues that we might never recover, most of these gaps are areas that can be addressed given more time, since we have a huge body of material (both specimens and research) that we can use to tackle these issues.

The central problem is perhaps the very one of support, with so little value given to this area of palaeontology despite its central importance in everything else, but very few academics do not maintain a research interest in anatomy, so it is an area that continues to push forwards even as it is marginalised. However, a resurgence here would surely help all other branches of the field and I hope most fervently that this is something that will happen sooner rather than later. If so, many of these spaces in our understanding would be addressed much faster and more easily than almost any others covered in these pages.

8

Mechanics and Movement

UNDERSTANDING HOW ANIMALS move in a mechanical sense, be it running, jumping, climbing, turning, grasping or biting, relies on a depth of knowledge about multiple aspects of their anatomy. Muscles move things, tendons hold muscles on bones, ligaments link bones to each other, and cartilage provides a cushion for joints, but perhaps the most critical of all are the bones themselves. Happily, this is an area of real strength for dinosaur research, and while the other aspects are important, many of them can be reconstructed in part from the bones to which they attach and by using living animals as models.

However, in the absence of living dinosaurs to study on treadmills, or dead dinosaurs to dissect in the anatomy lab, all of the work that can be done to elucidate their abilities comes primarily from generating computer-based models of bones, joints and other details and this requires a set of specialist knowledge combining maths, anatomy, palaeontology and computational skills. It is probably not a surprise, therefore, that there are only a few people working in the field of dinosaur biomechanics, though the numbers are growing rapidly and the digital tools available to them are similarly driving this research forwards at quite a rate.

Even so, while models may become ever more sophisticated and can be tested and verified with ever greater degrees of rigour, the vast majority of these analyses are focused on a few key aspects of biology (in particular running and the function of the arms in relation to the origins of flight) and are disproportionately applied to the theropods and to a lesser extent the sauropods, leaving some hefty gaps in our knowledge.

A weighty problem

Central to perhaps every aspect of mechanics, and almost every aspect of the biology of the dinosaurs, comes one issue: how heavy were they? This question has an obvious appeal in relation to dinosaurs, because of the huge sizes of so many of them, but size (or more correctly here, mass) is the basis of so much of the biology of any organism. How big you are determines how much you need of everything from food and water to iron and oxygen, how fast you can move, how much energy you burn, and the size of food items you can ingest. It can have a major effect on your success at breeding or controlling territories, how strong your bones need to be, and all manner of other intricate issues of scaling and resources. Put all of that together and it should be clear why so much effort has been expended on working out how heavy dinosaurs were.

The earliest estimates of dinosaur sizes were little more than guesswork. Numbers appeared in the scientific literature without any real calculations and early estimates of the weight of the large sauropod *Giraffatitan* ranged from 12 to nearly 80 tons – quite some difference of opinion! The first attempts with any rigour involved simple volumetric displacement; that is, you got a model of your dinosaur and dunked it in water to calculate the volume, and then scaled that to the density of a typical vertebrate.

Crude though this was, it had more than a little merit, however there were two big issues. The first was that we still didn't realise quite how pneumatic many dinosaurs were and often far too high densities were used, especially for the sauropods. The second one, which is the ongoing and not-quite-insurmountable problem, is that to get an accurate volume estimate, you must have a really good idea of what the dinosaur looked like when it was alive with its full complement of muscles and the rest.

Things have moved on considerably and we do now have a good idea of some detailed intricacies of dinosaurian body shape. The size, shape and positions of various scars on the bones show where muscle groups attach, and comparisons to birds and reptiles can help elucidate how large these were and work out what groups helped to perform which actions. Unusually well-preserved fossils give us an idea of how large the cartilage blocks were between the bones and key joints, giving a more accurate picture of the exact proportions of these and providing us with an opportunity to calculate how high the loading (and thus the weight) would have been on such joints.

Better understanding of the pneumatic nature of bones and how heavy they would have been, plus what other air sacs in the body must have connected to them, gives a much better impression of the volume of tissues present. Added to that, the very act of being able to put bones into a computer and move them around helps enormously with articulating ones such as ribs correctly, which can have a huge impact on the volume of the body.

Shove all of this together and it should be no surprise that we consider our current generation of dinosaur digital models to be far superior to those from even a decade or so ago, let alone the 1960s. More importantly, multiple different methods of estimating mass from numerous different research groups are all converging on a similar value for some species, and that huge variation seen in older estimates is vanishing.

Even so, some of these numbers remain quite varied. It's not uncommon to see *Tyrannosaurus rex* being reported as being anything from about 3 to 7 tons at adult size. How accurate are either of these, given that one is more than double the other? First off, not all of these estimates are based on the same specimens and some are bigger than others. Mass goes up quickly, even when you scale things up only a little, because it's increasing in all three dimensions, so something that is twice as long will be eight times heavier, since it's also twice as wide and twice as tall.

It's easy to look at two animals and think one is 'twice as big', but mass is the critical measure and that will be eight times more. So, even having an animal 10 per cent larger can be an increase of more than 33 per cent in mass, and thus the difference between a 5-ton *Tyrannosaurus rex* and one of 7 tons is not much to look at on a skeleton. Then there's the question of how stocky do you make an animal? Bones, muscles, viscera and lungs can all be factored in, but some individuals are fundamentally bulkier than others; they can simply be much broader, but they could also be more heavily muscled or carrying lots of fat.

Lots of animals change weight by huge margins over a year as they add and lose weight according to local conditions, so even a single individual animal could be anything from 4 to 6 tons over the course of a year. Even rigorous methods carried out on slightly different individuals, with one tending towards a little more svelte model and the other being a bit chunkier, would both be potentially correct. That means we're going to be left with ranges of numbers even for single individuals, and there can never really be a single 'correct' answer for a species.

Coupled with the range of sizes of so many dinosaurs where we do have a decent number of specimens, and the fact that there's not always an obvious plateau where they reach 'full' size in the way of mammals and birds, this is about as good as we are ever going to get.

The limits of large size

This still leaves open some really intriguing possibilities that have yet to come under serious scrutiny. First of all, just how big could some of these animals actually get? There are some potentially mechanical limits on size, and working out when dinosaurs might have hit them is difficult. As noted above, getting bigger means your mass goes up considerably and the bones will have to take that increase, meaning that larger animals have disproportionately large bones to support them and then need even larger muscles to move them, and will need to get oxygen into the body to support it and move it around.

Biological tissues have mechanical constraints and at some point if an animal keeps growing it will hit the limit of the heart to move blood or the lungs to take in oxygen, the ability to find enough food to feed such a huge system, or the muscles to move the animal around. The limit for sauropods has been suggested to be anything from 100 to 150 tons, which is a pretty broad range, and the upper estimates for the biggest sauropods are around 80 tons, which is quite close.

Exactly how large the big theropods could have been is another matter entirely, since they had only two legs to support their weight and they probably needed to be a bit more active to hunt than a sauropod did to catch its prey of leaves. That probably puts them in a massively lower bracket, but it could still be rather higher than the 7–8 tons often given as the upper ranges on *Tyrannosaurus*, the heaviest (if perhaps not quite the longest) theropod.

The very largest sauropods in particular are all known from extremely fragmentary remains and none of them are even close to being a complete skeleton. Each new discovery or analysis is generally hailed by the media as being 'the largest ever',* but first of all the variations and uncertainties described above mean that it's impossible to compare any of these animals properly with each other. It's pretty much a toss-up between all of the various species as to which of them may have been the biggest,† but in every case it's an argument built on less than half a skeleton and there's no real way to choose between them. Even then, these are single animals and that isn't going to tell you very much about the population as a whole, or what the big ones within that species may have been like.

A few rare medical conditions aside, tall humans typically top out around 2 metres tall (6 feet 6 inches), but there are people out there who are 2.15 metres (7 feet) or more, and unless you watch a lot of basketball you might never have seen anyone that big in your life. Even if you have, we recognise that these are truly exceptional people, representing a tiny fraction of the world's population and rarer than one in ten million.

That's true of animals, too – an annual survey of up to 20,000 alligators in part of the US throws up only a few animals over 4 metres long, even though the world record is over 4.5 metres. Of course, that record is still only the largest known and accurately recorded example from measurements taken in the last hundred years, on what is a

* Quite literally. Between writing that sentence and this book being published, three new sauropods were described, all of which were initially claimed by the media (if not the researchers themselves) as being the largest ever.

† By 'biggest', scientists really mean 'heaviest'. The mass of an animal is linked to almost every major part of their biology and so is the key concept behind any discussion of size. Things like very long necks or tails could produce animals that were very long or tall without actually being all that heavy.

stressed and depleted population. The chances that even this whopper of an alligator is close to representing the largest one ever is vanishingly small.

So when we talk about this dinosaur or that as being the largest ever, it is going to be very shy of the true figure. For the larger dinosaurs, we have only a handful of good specimens that would allow us a rigorous reconstruction of them and give a good mass estimate. *Tyrannosaurus* is a case in point, where there are perhaps a dozen good skeletons, and as a focus for a lot of research some work has been done on various different individuals. But that's a dozen of them. An equivalent to finding a 4-metre alligator might take centuries of excavations to produce hundreds of skeletons and have even a chance of finding it, let alone getting near that 4.5-metre record, or the ones above that.

Cutting a very long story short, the idea that we have any dinosaur specimens even close to being the largest of a given species is all but impossible. More than that, however huge our specimens are, they are probably very, very short of the kinds of sizes of animal that a species might produce on occasion.

That puts out there the possibility of animals much closer to those mechanical limits than we are seeing from the fossils we have. Sadly, we will probably never find a sauropod skeleton that we can convincingly restore as being over 100 tons, or a 20-ton theropod, because these would be one in a million, or one in a billion animals. But when you factor in the possible extremes of size for species, I think these are surprisingly plausible.

Given further evidence of the distributions of body sizes within populations of dinosaurs, coupled with better sample sizes, and we might at least be able to predict what kind of statistical tail is reasonable and project how large those animals could have been. More work on how various tissues and systems scale up can also inform us if that tail would be approaching those potential mechanical limits of bone and muscle, and if our dinosaurs reached the theoretical biological maximum. It would seem extraordinary that animals ever evolved to the point that physics was the limiting factor, but the presence of the super-giant sauropods does seem to raise that possibility.

Standing tall

Mass is therefore pretty crucial to working out how animals function as biological machines, but even working out how they stand has its issues. For far too long, dinosaurs were reconstructed as simply giant lizards, despite some stark and clear evidence to the contrary, not least the shape of their femora, meaning they must have had an upright posture. Even so, their reptilian nature meant that palaeontologists clung on to various notions about them based on their living relatives.

One of these that survived pretty much into the 1980s and still turns up occasionally in the lower end of dinosaur books is the idea that their tails dragged on the ground in a very lizard-like manner. This seems to be a combination of cheap books endlessly recycling old text and images that represent out-of-date science, along with people not appreciating how much the field has moved on, and assuming that what they encountered in their childhood holds good decades later.

In this case, we know that tails did not drag, because their muscles are arranged such that the tail would be held off the ground, and in the case of the ornithischians, the spine also had numerous ossified tendons gluing the bones together, which would have made the tail relatively stiff and straight. Even more convincing are the endless trackways for dinosaurs without any kinds of marks where the tail would have been on the ground. A few are known, but these are exceptionally rare and show only the tip of the tail hitting the ground, and although it's inevitable this would have happened occasionally, it certainly wasn't normal.

Moving to the other end, the posture of the neck has also been contentious in recent years, especially for the sauropods. For a long time, these were reconstructed in an S-shape with the head held high, but in the 1990s and early 2000s there was a move to bring this down and have the long neck held out straight in front of the animal, as a counterbalance to the tail being held out straight behind. This was supported by the joints between the vertebrae in the neck aligning such that a straight neck would maximise the overlap between the little connecting projections from the bones and help support the weight of the neck.

However, most recently it's been shown that terrestrial vertebrates as a whole normally hold their neck in an S-shape, despite this reducing the overlap in those supporting projections. As a result, the S is returning, but there is still much more to be done here and there may well be some exceptions. For example, as I mentioned earlier, the Chinese sauropod *Mamenchisaurus* is far from the largest of sauropods and yet it has a neck a staggering 13 metres in length, fully half the total length of the animal (including the tail!), and also shows some greatly elongated and dead straight ribs on the neck vertebrae, which must have had some role in stiffening and support and suggests a relatively straight neck in these at least.

Even the fundamental posture of many dinosaurs is somewhat uncertain. It looks increasingly as though the hadrosaurs were essentially quadrupedal, despite having long been considered almost kangaroo-like in walking on all fours when moving slowly, but at high speeds transitioning to using only their hind legs when running. Others are apparently more complex, though, with some of the forerunners of sauropods starting out life on all fours as hatchlings, and then becoming bipeds as they grew. Quite how they did this or even why hasn't been looked at in any detail, and there would have needed to be some interesting rearrangements of both the arms and legs as these animals grew in order to shift from one to the other.

As with so many things, this would also have other interesting knock-on effects for what kind of food they were accessing at different heights, and how they could have moved both before, after and during the transition phase. This pattern is still more complex, since the ancestors of the prosauropods were bipeds and the sauropods themselves were quadrupeds, so at some point we have a bipedal lineage shifting where the juveniles are quadrupeds and the adults bipeds, but also a transition from bipedal animals to the fully quadrupedal sauropods. That's a lot to sort out and the increasing number of studies on the prosauropods, coupled with the more embryos, hatchlings and juveniles being discovered, provide excellent fuel to the fire of understanding the biomechanical changes that would have taken place.

We do lack any kind of obvious transitional sauropod, and their fossil record into the Late Triassic when they first appeared is not very good, but there is a lot that is likely to come in this area in the near future as new specimens are uncovered, and new mechanical analyses of existing specimens come to the fore. Perhaps this is another case of the retention of juvenile characteristics leading to a radiation of new forms, but at the opposite end of the scale to theropods shrinking down to birds, with hatchling prosauropods in essence going on to produce giant sauropods.

Despite the ongoing discussion over possible bipedal and quadrupedal postures for various dinosaurs, one thing we are near certain of is that they could all rear up onto their hind legs for the simple reason that without being able to do so, mating was probably impossible for them. I'll deal with the issues of dinosaur reproduction in more detail in a later chapter, but this act at least must have been possible for most.

That in itself is astounding when you consider the size of the largest sauropods and that males weighing perhaps 80 tons would need to rear up and mount females, which would then have been taking much of the male's weight on their back. The loads would be quite phenomenal and the ramifications important. There has been some important biomechanical work on rearing in dinosaurs, but to date little at the upper end of sizes and what this would have meant for their flexibility, balance and support.

Getting around

All of this is still essentially a form of static mechanics and, of course, dinosaurs could move. There has long been intensive research into the basics of walking and running of dinosaurs – how they actually put one foot in front of the other, and in particular what kinds of speeds they might have been reaching. In recent years this has also encompassed more technical working into things such as their ability to accelerate and turn, to give us an idea of their agility too.

As with body mass estimates, these studies are often a bit light on hard numbers, since there are so many variables that are uncertain or have a wide range of errors associated with them, but it is possible to do good comparisons between species based on these analyses. So as a result, it might be hard to say whether or not an adult *Tyrannosaurus* could hit 30 kph (almost 20 mph), but we can say with confidence that their juveniles were fast and could turn better than could the adults. Even here, though, to date this kind of work has been limited in the number of species that have been analysed and so even comparisons are currently rather limited.

It would be nice to see how well a large *Tyrannosaurus*, for example, might match up in terms of acceleration, top speed, endurance and agility against classic potential prey species like *Triceratops*, and how that might change for each species as they grew. It is probably only now a matter of time until sufficient studies have been completed on enough species that we can start to make really meaningful comparisons between various different species, looking at their relative abilities to catch each other (or escape) and for major activities such as long-distance migrations.

The origins of flight in birds remain a massive area of research and also the possibilities of gliding and flapping in various animals. We have a pretty good idea of the basic issues with the mass of the animals in question; their anatomical arrangements with how they held out their different wings and used them as control surfaces (the arms, legs and tail), and the amount of lift being generated. Progress remains strong here and so while there are areas that need to be filled in and more to come, it's not an area that is either under-researched or likely to produce major breakthroughs, since most of these are, if not in hand, certainly pointing strongly towards the answers. However, one of the two major areas that does still need work in terms of the origins of flight is the transition from gliding to flapping.

Powered flight has evolved only four times: in insects, bats, pterosaurs and the birds.* Insects are not much of a help with bird flight; the origins of pterosaurs (the flying reptiles that lived alongside dinosaurs) remain poorly understood generally and even less so in terms of their development of flight; and fossil bats are extremely rare. So all the practical research is on the birds themselves and their dinosaurian ancestors, and while the information we have from the fossil record is now exceptional and detailed, with dozens of species known from

---

* Though several recent studies have suggested that in fact powered flight might have evolved multiple times, with several of the small and feathered dinosaur groups close to the birds also independently getting into the air.

hundreds or thousands of specimens between them, the transition from gliding to flapping is poorly understood.

Ongoing discoveries from China in particular, where there are numerous feathered dinosaurs apparently from the very point at which flapping emerged, are likely to help fill in this gap soon, but there remains another area that also needs more work.

In order to fly in some way, animals need to launch and become airborne. The vast majority of birds get most of their launch power from their legs, so even if their wings are pumping hard, it's their running and/or jumping that gets them aloft. Far too little emphasis has been placed on this important aspect of bird flight (by both palaeontologists and science communicators), and the jumping capacity of various small dinosaurs needs to be investigated to see how this lines up with apparent capacity for flight, since there should be a fairly strong relationship between them. That said, perhaps jumping was a precursor to flight and it would be intriguing to see how far back into various lineages the ability to jump well might appear. And as for the larger theropods in particular, could they even jump?

A related issue in terms of launch is the ability to climb up vertical surfaces, most obviously trees. Various studies have strongly advocated small bird-like theropods as being capable climbers, but also as being very unlikely to have engaged in this behaviour. Both clearly cannot be true, and as ever the truth may lie somewhere in between, since some animals don't appear to have many obvious adaptations for climbing and yet are surprisingly good at it, as seen by the number of goats that end up in trees.*

If flight did evolve from fundamentally terrestrial non-climbing animals that would be a bit of a surprise, and so establishing their abilities in trees is an important component of working out where these animals were launching from and by extension what evolutionary pressures were triggering flight, since there's a huge difference in terms of getting an initial elevation from the ground or from a tree branch.

---

* Quite seriously, if you have not seen this phenomenon, it is worth looking up online. Goats in Morocco in particular are very good at climbing up into trees and out onto surprisingly flimsy branches.

Certainly it would seem that many of the little gliders must have been climbing, or how else would they get to a high point when gliding can only take you down, but that doesn't mean the first flappers were necessarily taking this approach. In addition, the little theropods are not the only dinosaurs that might have been making it up into the trees, and there are plenty of small ornithischians that were light and agile and with somewhat grasping hands.

Admittedly, none of them look too well suited to an arboreal life, given in particular their lack of classic climbing hallmarks such as well-developed claws for gripping, or arms with a wide range of movement to reach awkward branches. But again, this doesn't rule out them being able to get into and around some trees, and this is an area where all the work has focused on the origins of birds and with almost none on any other groups.

Dinosaurs were primarily terrestrial animals and we've commented here on the issues surrounding flight, but sooner or later most species would have encountered large bodies of water. As noted in an earlier chapter, dinosaurs seem to be oddly bereft of semi-aquatic or even highly aquatic species, but that doesn't mean that the others didn't or couldn't swim. Indeed, while some excellent work has been done on how various dinosaurs would float given their lung volumes and other pneumatic bits, this doesn't say much about how well they might have got around when in water.

Few may have been good swimmers (though even the credible candidates for a semi-aquatic life have had no in-depth analyses of their likely forms of locomotion), but it would certainly be interesting to know which species could have swum well and which would have floundered. Numerous bonebeds of larger ceratopsians are known where the animals appear to have drowned, but we don't know if these were relatively capable swimmers overcome by floodwaters or unexpectedly difficult crossings, or animals that could barely make it to shore at the best of times, or even if mass drownings might have been a fairly regular occurrence.

One final terrestrial aspect that's been little studied is that of digging. Many dinosaurs likely dug holes in order to create nests in which to lay their eggs, and while the giant claws on the hands of some sauropods have been linked to excavating nests, few others have

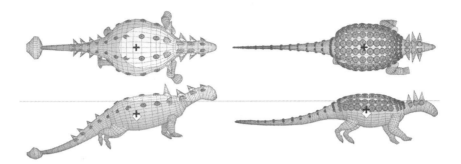

Digital models of a pair of anyklosaurs floating in water (left,
Euoplocephalus, right *Sauropelta*). Models like this
allow us to assess how such animals might have floated,
how they might have swum, and what could happen
to their bodies if they died in water. Images by
Don Henderson.

had anything like the attention. The little insectivorous alvarez-saurs, for example, have been shown to be well adapted to making holes to uncover their food, and the little ornithischian *Oryctodromeus* was apparently living in burrows and also shows some adaptations to digging, but there has been little systematic study of the mechanics of their potential to dig, let alone of the host of other dinosaurs that in some way were probably capable of excavating soil.

As with the point about goats above, animals don't necessarily need to show a suite of adaptations to be decent and active diggers in some circumstances. The (orange and rather furry) Sumatran rhinoceros can dig quite large and deep pits to make into a mud wallow for themselves, despite lacking any real specialisations for the activity. Though for the rhinos this is probably helped by the relatively soft and moist rainforest soils, it would seem odd indeed if no other dinosaurs were engaging in similar behaviours and, when studied, some could easily reveal something of an anatomical link to doing just this kind of thing.

Overall, biomechanics is a booming area of research for dinosaurs, with the ongoing advancements in technology allowing better and easier analyses of complex motions and stresses and enabling researchers to delve deep into the physics of posture and motion. Of all the aspects covered in this book, this is perhaps the one field where most of the major pieces are in play and it's mostly a matter of time until enough researchers tackle these questions and then we will have major answers.

In the background, work by non-palaeontologists developing new technologies and software to allow better and faster calculations, and anatomists providing more in-depth understanding of living animals, will keep pushing this field into being ever more accurate. For now the overlap between palaeontology and biomechanics is limited, in that there are few researchers with the expertise to dive deep into this area, but it's growing in depth and breadth and this is surely an area with some major new innovations and answers coming in the very near future.

9

Physiology

O NE OF THE questions asked most often of any dinosaur researcher is: 'Were they hot or cold blooded?' The very short answer is, 'It was probably different for different dinosaurs, but we don't really know that much' – and the much longer answer is the rest of this chapter. As with so many topics, we do know quite a bit about dinosaur physiology, even if what we know tackles only the margins of some of the problems or gives certain key answers, though without necessarily a good understanding of the underlying reason behind them. To begin, however, we need to get to grips with the basic issue of that apparent dichotomy of hot versus cold blooded.

As you may have already known or suspected, this is a false dichotomy and one that also masks a range of differing conditions between various living animals and by extension animals like the dinosaurs, too. Imagine a simple graph with two axes, and the first axis indicates the typical internal temperature at which an animal operates, and can vary from cold to hot. The second axis is the degree of variation in that temperature. Animals will sit on various parts of this graph normally, but also plenty of animals also move around the graph depending on various factors such as the time of year or how large or how active they are. So rather than a simple 'hot or cold', we have 'varying degrees of hot or cold' and 'stays warm or stays cold or varies', and things moving between different bits of this.

Humans (and most mammals and birds) are described as homeothermic endotherms, meaning that we maintain a single temperature and it's driven by our own internal physiology. If it's cold, we'll burn more energy to make sure we stay warm. Most reptiles would be described as the opposite of this, heterothermic ectotherms, meaning that their internal temperature varies and they don't normally generate much of their own heat, but need to draw it from the environment. This is why there are so few reptiles in cold climates (despite there being many mammals and birds) and we are used to seeing snakes and lizards basking in the sun to pick up heat. These are the best-known versions and thus the reason for the commonly understood hot versus cold dynamic, but there are numerous alternate strategies to this, which dinosaurs could, and in some cases almost certainly did, employ.

First off are the homeotherms that are, well, not necessarily that homeothermic. We are familiar with the idea that some mammals such as squirrels and bears hibernate and see off the worst of the winter by simply sleeping through it. When they do this, their physiology drops and their energy burn goes down considerably. True, they are not very active and are staying warm and well above the local environmental temperature, but equally they are not maintaining their usual high temperatures, so are certainly showing some strong variation in their temperatures.

This is not solely a strategy of hibernating animals, though; some others like dormice and hummingbirds will essentially do this every night, going into a state called torpor that is essentially a mini-hibernation of a few hours to preserve their energy. There's no reason at all why dinosaurs could not have done something like this and it would be very hard if not impossible to detect from fossils. It's also worth noting that not all mammals are very good at maintaining their temperatures and various species from warm climates really struggle if the environment gets too cold – platypus, naked mole-rats and some lemurs are all reliant on local heat to stay at a healthy temperature and keep warm.

Second are those animals that cheat their way to higher temperatures. As organisms grow, their surface-area-to-volume ratios change and large animals have proportionally low surface and high volume even if they maintain an identical shape. A cube with sides of two units will have six sides of four square units for twenty-four units, and a volume of two by two by two for a volume of eight units cubed. But a cube of sides twenty units long will have sides totalling 2,400 units and a volume of 8,000 units cubed. The second cube might be ten times bigger, but it now has much more volume compared to its surface area than the small cube. This means that any little heat generated by various processes like digestion or from moving around will not be lost easily by large animals, since they have a high volume of tissues making that heat and not much area to lose it.

Thus large animals can potentially maintain a high body temperature (and so can be called homeotherms), despite not generating their own heat directly with specialised physiological processes (and so essentially being ectotherms). Some large sharks and tuna do this and so too can the larger crocodiles, essentially maintaining a high temperature through being large, a condition called gigantohomeothermy. It's not quite that simple, admittedly; the fish do this in part by constant swimming to generate muscle heat, and the crocs can help stay warm by taking to the water, which at night is much warmer than the air, but equally, even the largest of these animals are a fraction of the size of the largest dinosaurs, where the effects would be exaggerated.

Hot dinosaurs?

So, on to the dinosaurs themselves. One central part of understanding dinosaur physiology is their growth rate, since this is something we can measure in detail. Most vertebrates as they grow lay down layers of bone onto their skeletons. When growth is fast, this can be relatively low density and when growth slows, this tends to be denser. Cut through a bone and put a sliver of this under a microscope and these differences show up as rings in the bone. Just like a tree, which has the same rings for the same reason, the dense and darker rings represent periods of slower growth that essentially means the winter months. So, each ring generally indicates a period of slow growth that occurs annually and therefore one ring represents one year.

Bone constantly changes and is being broken down and built back up, so these rings can be obliterated over time, but if you have multiple members of a species, you can count the rings up to a certain size and then switch to a larger animal, carry over the count from the smaller one and keep going. Such a method is a little crude at times and it can be misleading, with events like a drought leading to the possibility of slow growth and an extra ring, or a very warm winter meaning one is missing, but on average it should be a good indicator of the age of a given individual. Couple this with any indication of the size of the animal from which you have your samples and you can reconstruct how quickly they grew, looking at how much weight they put on per year.

Pretty much every dinosaur that has been examined appears to have a fast growth rate comparable to that seen in modern mammals and birds. This can only be achieved by having a high body temperature and maintaining it at a high level, since you need to be warm and active to gather lots of food and put that into growing quickly. Even various ectotherms can't usually maintain such growth rates when kept in hothouse conditions, so it's not simply a function of being in a warm place.*

So at some level, many, perhaps even all, dinosaurs were 'warm' and the smaller species can only have been maintaining this at least partly through their own physiology keeping them warm. There are also numerous dinosaurs the growth of which palaeontologists have yet been able to measure their growth and while it's notable that we have a uniform result from those we have assessed, and they come from multiple different lineages and different times and from differing environments, it leaves a lot of scope for various exceptions being out there that we have simply not yet analysed.

For those that lived in hot conditions (and remember huge swathes of the planet would have counted as being hot for much of the Mesozoic), they could have been like lemurs in the sense that they were providing some internal heat, but might have struggled if the temperatures were lower, in which case they would have been some kind of halfway house towards what we might call true endothermy. On the other hand, while some lizards and snakes can do alright in cold climates (provided they can stay underground for the worst of the weather), crocodiles generally cannot survive in colder climates.

---

* There is a caveat to this in that with high enough temperatures, even things like alligators that normally lay down growth rings in their bones don't seem to have them. This means that it is possible we are undercounting the age of the dinosaurs and by extension increasing their apparent growth rates. There are some good reasons to think that this isn't a big problem for our estimates overall, but for some individuals it might cause us to come up with some spurious numbers.

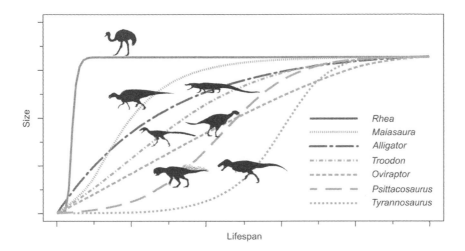

	Rhea
	Maiasaura
	Alligator
	Troodon
	Oviraptor
	Psittacosaurus
	Tyrannosaurus

A graph of the reconstructed growth rates for
a variety of dinosaurs and both a modern rhea
and alligator. Image by Jordan Mallon.

American alligators, perhaps surprisingly, can hibernate by staying
in frozen ponds with only the snout above the water, but this would
hardly be an option for most dinosaurs. It's reasonable to infer that any
of them that were residents in especially cold climates were likely
producing some heat, or they simply would not have been able to
survive.

Some have argued that anything with a good covering of filaments
or feathers would have been endothermic, and thus any of the
maniraptoran theropods and the early birds would have had physiologies like those of modern birds. This may be correct, but it already
overlooks various mammals and birds with full coverings that are not
full homeothermic endotherms and it also misses out a few odder
animals. Moths are heterothermic ectotherms and many have very
thick coats of insulating fibres on them. Operating at night, they can

end up struggling to fly unless they stay warm and they get their heat from using their flight muscles. You might well have seen them sitting still but trembling, and this is them buzzing their wings to warm up before they can fly.

Obviously there is an enormous evolutionary gap between them and the dinosaurs, and therefore they are a poor model in lots of ways, but the central point is that even animals with very high energetic demands for behaviours such as flight can still be both full ectotherms and end up with insulating cover. Early birds don't appear to have been very good fliers either, so while flight is very expensive energetically, if they couldn't do it much at all, there would not be any need for a sustained high level of physiology. They may have possessed this for some other reasons, but the idea that feathers and flight must somehow link to homeothermic endothermy is incorrect.

At the other end of the scale, there are large and apparently fully feathered theropods that might have struggled with overheating. There is a huge question over the possible plumage or otherwise of large tyrannosaurs (which will be covered in the next chapter), but other theropods that probably exceeded a ton were probably covered in feathers, and at least some lived in warm environments, too. The largest modern mammals and birds that live in hot climates tend to have reduced hair or feathers (the bald heads of vultures, the bare legs of ostriches, the limited hair on elephants and most rhinos), because the issue is much more about trying not to overheat than it is to stay warm.

Large dinosaurs could have responded in the same way and reduced their insulation to stay cool, provided that they were endotherms in the first place, of course. Recent studies have shown that the arrangement of tissues on the top of the heads of several dinosaurs (especially large tyrannosaurs) would have functioned to help cool down the brain, and this applies even to species that lived in relatively cool climates, and again points to the idea that these animals were fundamentally hot at some level if they required such features. Other strategies could have been employed too, such as being active primarily at night, or using water or shade to help cool off.

Elephants have large ears to help shed heat, so there could have been soft tissue structures such as dewlaps or other folds of skin present in some dinosaurs to help them avoid overheating. Remember too that camels are large endotherms that live in deserts, but maintain thick fur to keep them warm at night and in winter, when there can be a lot of snow around. So again, in addition to the complex (and unknown) physiologies of dinosaurs, there are various alternate solutions that we may not be able to detect.

Too big and too hot?

Sticking with large size, the sauropods provide perhaps the most intriguing issue when it comes to being big and what that may mean for their physiology. The points made above about overheating would absolutely apply to sauropods and, even without any kind of insulating cover, some would potentially have had overheating issues and would need anatomical or behavioural adaptations to overcome them. The long necks and tails would certainly have helped provide some extra surface area to keep them relatively cool, but it may not have been enough.

The real issue comes back to gigantohomeothermy, whereby animals essentially stay warm by being big. The very largest sauropods are at a minimum in the realm of 50 tons and probably quite a bit more; that is an entire herd (if a small one) of elephants bound up in a single body. Elephants can have trouble avoiding overheating when the temperatures go up, so that strongly implies that any large sauropod, or to a degree, perhaps any large dinosaur – though it would be more extreme in sauropods and thus easier to make the point – would almost have to be an ectotherm. If they constantly produced their own physiological heat, they would end up cooking themselves (perhaps almost literally) if they had to move around in the sun.

That may settle the condition of adult sauropods, but not the juveniles. A hatchling sauropod for even one of the larger species was probably only a few kilos in weight, the size of a typical house cat. That means they would go through something like a 20,000-fold increase in size to reach the size of a 50-plus-ton adult. Certainly no hatching sauropod or even any dinosaur under a ton would be capable of gigantohomeothermy and yet they still show evidence of that rapid growth rate.

That leads to a single conclusion: the youngest sauropods would need to have been operating some form of homeothermy when young and then potentially have lost this at some point in their development. I'm not aware of any other organism that undergoes such a shift (though there are few living homeotherms out there that get to even a ton, so perhaps it's not surprising), but it would be a quite incredible feature of their biology. Of course, now come the obvious caveats and issues with this, which are that we don't know quite when this change would have hit in terms of their size, or what transitions they may have gone through with all kinds of variables of their age and habits and habitats, but some major physiological shifts must have happened.

In the previous chapter, I mentioned the possible adaptations that might be present in various herbivorous dinosaurs to allow them to digest various plants better. Features like a divided stomach would make things more efficient, but a key part of this would also be their physiology. If, as surmised above, many dinosaurs were constantly warm, this would increase the activity of enzymes and allow the bacteria breaking down plants in their gut to be more effective. That would return them more energy faster from whatever they ate, but for those who were endotherms, they were also burning more energy to stay warm.

A classic argument about dinosaur physiology in the 1980s was the relative abundance of predator and prey species of dinosaur in the fossil record. Endotherms with their high metabolism need more food than do ectotherms, so, quite simply, if dinosaurs were 'hot' there should be relatively few carnivorous theropods and if they were cold, there should be many more of them. Such an analysis suggested that they were warmer rather than cooler, but there are so many caveats (population structures, growth rates, differential habitats, preservation and collection biases) that the result may be all but meaningless.

Still, the central point is quite correct: endotherms need more food and need to be more active to keep up with their own physiological demands. That makes things both more interesting and complicated with the potential switches from starting with internal heating to losing it later as dinosaurs get bigger. There might be more dinosaur there, but even if they were maintaining a high temperature and high activity levels through gigantohomeothermy, they were still not burning fuel solely for heat, so they should have been eating proportionally less than the much smaller members of their species.

If this is the case, then things start becoming increasingly complex with trying to work out exactly how much food a dinosaur might need to survive and how much a herbivore could be sustained by the local flora, and thus, by extension, how many carnivores were out there.

This is another area that hopefully we can begin to delve into in more detail as our understanding of the physiology and temperature controls of living animals improve and we get a better handle on the mass estimates and growth rates of dinosaurs, as well as the local temperatures and conditions they lived in. Pulling together these kinds of data should allow us to delve into the problems with greater accuracy, and as they are improved (if not actually resolved), these will have major knock-on effects for our understanding of dinosaur ecology and interactions between various species, and how ecosystems as a whole that were occupied by such large animals might have functioned.

Damage and healing

One other area of dinosaur physiology we do know something about is their response to injuries and infections. Various dinosaurs show evidence of injuries and infections and even illnesses that we would recognise, such as arthritis and cancers. There is only so much we can determine from these, especially when it comes to diseases and infections where there could be any number of pathogens from bacteria, fungi and viruses, although some of the patterns present are revealing. Rather like modern birds and reptiles, dinosaurs appear to have been capable of keeping infections localised.

Even major infections that left marked swellings and deformities on bones tend to be in single spots and don't spread around the rest of the bone or extend out onto others, as is all too often the case for infections in mammals that reach the bone. Similar to reptiles, but perhaps somewhat less like modern birds, the dinosaurs were also capable of healing incredibly traumatic injuries. There are various dinosaurs out there with significant parts of the tail missing and partial arms, in at least some cases apparently bitten off by theropods, but showing signs of extensive healing, with the animal having survived the injuries and continued to live a long life afterwards.

While mammals can and do survive similar injuries, it is more unusual and suggests that the dinosaurs, in this way at least, were rather reptile-like and resilient to major traumas.

There is also a related aspect of dinosaur biology that has caused much discussion and is related to injuries and healing, and that is their potential ability to inflict damage to others through venom. In addition to the entirely fictional idea portrayed in the film *Jurassic Park* that *Dilophosaurus* could spit a blinding venom at other animals (and, as necessary, treacherous employees), the small feathered theropod *Sinornithosaurus* was once suggested to have had a venomous bite. The evidence put forward for this was rather weak and few if any palaeontologists have accepted the assessment, but that does not rule out anything like this in other species.

A wonderful skeleton of the hadrosaur *Parasaurolophus* with crushed and healed spine forming a deep notch in the back. Photo by David Evans.

While complex venom delivery like that of a snake would be pretty obvious due to the modifications necessary to the teeth, various reptiles have a saliva that contains toxic compounds and the mere act of biting something can introduce these to the bloodstream without the need for grooved or hollow fangs. Short of that kind of evidence, though, it's going to be very hard to provide confirmation for anything like this in dinosaurs. Snakes aside, it is already pretty rare in reptiles generally and not currently known in any crocodilian or bird, so there's no immediate reason to think that many or perhaps any dinosaurs were venomous, but it's certainly a possibility and one that remains frustratingly out of reach to assess.

Physiology is an enormously complex subject and here we have covered only a few limited aspects of it for which we have some good evidence and understanding. Clearly there is scope for serious improvement, but given that we don't even understand the full nuances of human physiology, let alone that of most other living animals, it would be an impossible task in a dinosaur.

That said, given the profound limitations of trying to assess biochemical processes and their complex interactions across the biology of an animal even when it is alive, the amount of detail that we are starting to uncover for dinosaurs that have been extinct for 65 million years is quite remarkable. In particular, the areas of how dinosaur bones respond to injury are likely to develop the fastest, as these are a ready source of information with limited ambiguity in a lot of cases. Our ability to examine these at a near molecular level with modern scanning techniques will yield some important new discoveries, and help complete the gaps in our understanding of where dinosaurs sit in their biology between the reptiles and birds.

10

Coverings

ALTHOUGH FOSSILS OF soft tissues are very rare compared to bones and teeth, among the most common of those that do preserve are the various parts of dinosaur skin. Skin (especially scaly skin) is a much tougher structure than the internal organs, since it is there to protect the owner from the elements, so it's perhaps not a surprise that we have quite a lot of it.

The skin and its component structures (better called the integument) are the barrier between an animal and its environment and play a major role in basics such as regulating heat and keeping out the elements, or nasty things like diseases and the teeth of predators, and its colour, texture and composition can have all manner of roles in communication. The integument is therefore a major component of dinosaur biology and yet there is much to work out here.

The basic appearance of the dinosaurs was arrived at quickly by the original discoverers of the dinosaurs: they were large reptiles and would have had scaly skin. These fundamentals were soon supported by various finds of fossil skins (or at least their impressions), but the discovery at the end of the last century of feathered dinosaurs opened up a huge realm of new possibilities. Since then, ever more numerous discoveries of dinosaurs with odd kinds of scales, filaments, fuzz, horns, and feathers has made this once simple field (all scales) much more complicated and confusing, but also more intriguing and interesting.

Some of the earliest illustrations of dinosaurs show them as having lizard-like scales, even at a time when they were known from a handful of bones and before any fossilised skin had been recovered. These went beyond the usual polygonal scales that make up most of the skin of reptiles, early illustrations often also including a tall row of triangular scales, which rose up along the spine in a rather saw-toothed effect.

These are present on many of the larger iguanas (perhaps again the early influence through *Iguanodon* shows up) and this feature has also become typical of many fictional reptilian beasts as well. Larger dinosaurs like the sauropods were usually shown with elephantine skins of thick hides full of folds and wrinkles, while armoured dinosaurs such as the ankylosaurs were shown with large scales over their spines and plates, in the same way that modern crocodiles have scales over their bony armour.

Such simple comparisons and extrapolations were more than reasonable at the time and were still acceptable right into the 1990s. By then, of course, a huge amount of information had been accumulated about dinosaur skin and appearances, and this supported fairly well the various reptilian morphologies that were illustrated by the Victorians 150 years earlier. Huge numbers of footprints are known for various dinosaurs, and while few show fine details, there are some that preserve clear impressions of scales on the feet and also the claws of the toes. Skin impressions are known, too, either where animals lay down in mud (leaving essentially what is a 'footprint' but of a flank, or neck, or tail), or where an animal died and the sediments that buried the skeleton also preserved some of the impression of the skin of the creature.

Such fossil remains gave us some pretty big hints about what kind of coverings dinosaurs had and how they looked. It supported the idea that they were reptiles and rather lizard-like ones at that, though it's rather more complicated than this.

Scales

Surprisingly common are actual parts of fossilised skin – the original biological tissue having turned to stone, as opposed to merely having left an impression of it. Skin is known for a variety of dinosaurs, though it is especially common for the hadrosaurs of the Late Cretaceous, and numerous examples of these have been recovered, generally associated with skeletons. Many remains of skin are little more than patches of only a few centimetres across, but occasionally skin appears in great swathes and whole limbs may be covered, or even most of the body recovered.

More than a few such finds have been described as 'mummies,' where almost the entire skin has shrink-wrapped around the skeleton and so the whole animal is preserved, including traces of muscles (and potentially internal organs), if in a rather desiccated state. In these cases, presumably the animals died in the open under a hot sun and shrivelled up before any large scavengers could get to them, or for that matter any bacteria or insects. Such carcasses are not uncommon today, especially in deserts, so there is a good model among living animals as to how skin can preserve under the right conditions.

These various examples clearly show that dinosaurs had different shapes of scales on different parts of the body. While there is some major variation between dinosaur groups, as a broad generalisation most dinosaurs were covered in a network of fine scales, often only a few millimetres across, even in the largest giants. Some, including several of the biggest hadrosaurs, also had much larger scales in rows in various places, often including circular ones that can be 10 centimetres across. There are also mainly smaller scales in the various joints, which would add a degree of flexibility to the skin of the elbows, knees and toes.

Those Armoured and plated dinosaurs often did have large scales that sat on the little armour pieces (these scales are sometimes called 'scutes', a word incorrectly used for the underlying bony pieces, which are properly called 'osteoderms' – literally meaning 'bone-skin'). Larger bones and pieces that stuck out, such as the plates and shoulder spines of stegosaurs, the horns of the ceratopsians, domes of pachycephalosaurs, and various spikes and knobs such as iguanodontid thumb spikes, would have been covered in a thicker layer of keratin. This tough material makes up skin, hair, fingernails, claws and feathers, as well as eyelashes, and is therefore a very versatile suite of proteins.

The skin or its components of all vertebrates also has a lot of keratin in it, but here we're concerned with the bigger units such as scales. Occasionally chunks of keratin are preserved around distinctive anatomical features of dinosaurs like horns and plates. As such, we know that these bony structures were covered in this way, just as cow and sheep horns are today.

Some of these can also reveal some novelties. Dinosaurs are best illustrated with a zigzag row of spines along their back, so that they look like a child's drawing of a mountain range, but at least a few species did have something like this. Only relatively recently were some small and triangular bones found associated with a specimen of the famous *Diplodocus*, which shows it did have a line of little spikes running along its back, quite possibly from the back of the head to near the tip of the tail. This is not likely to be a one-off and many sauropods could have had something similar.

One other area covered by keratin that is important are the claws. More correctly, the last bones of the hand or foot are called unguals and these bear the keratinous claws.* Ungual and claw shape vary dramatically between species and occasionally between fingers on a single manus or pes. Theropods typically have pointed, curved and narrow claws on the hand, which would be sharp, with more rounded and flattened ones on the feet. The early sauropodomorphs tended to have quite large and spiky claws on hands and feet, with the later sauropods having one large claw on each limb, and much more rounded and thin claws on the other fingers and toes. The ornithischians generally had something close to a miniature of a horse's hoof on each digit, though there were some exceptions, like the thumb spikes of the iguanodontids.

We do occasionally get claw sheathes preserved alongside the bony unguals of dinosaurs, generally when other things such as beaks or feathers are also preserved, and these have some rather conflicting patterns in their relationships. On the one hand, claws broadly match the unguals such that a very curved bone will hold a very curved keratin sheath, and long ungula will have a longer claw than that of a short one. On the other hand, the relationship between the two is not that close. Some claws greatly exaggerate the curve of an ungula or greatly extend the length of the claw as a whole, such that the shape change and proportions are quite extreme.

The raptorial claws of dromaeosaurs, for example, are already quite curved, but then the sheath can mean that this is more like a semi-circle in its arc and nearly points back on itself, a far greater curve than

---

* Well, usually. Humans and other primates are rather unusual in having nails that only partially cover the end of a finger or toe, rather than a claw, hoof or similar sheath that covers the whole of the end.

is seen in the bone alone and hard to predict from it. Thus, while it is clear that some therizinosaurs (the giant-clawed, 'scythe lizards') had truly colossal claws and the tyrannosaurs less so, the degrees and differences in all manner of species may be greater or less than we imagine. Such patterns are also true in living species and so this is an area that we may not be able to refine much – getting an accurate and detailed picture of claw shape may ultimately rely on specimens with the sheaths preserved.

In particular, we do not know much about how the scales of the faces of the various dinosaurs were constructed. It is common to see dinosaurs illustrated with big square scales around the eyes, nostrils and especially around the margin of the mouth. This seems to have become something of a trope (or meme, take your choice) based around the nature of the jaws of some of the larger lizards, with Komodo dragons being a great example of this. But this is hardly a universal in the reptiles, and such a feature may not have been present in any dinosaurs, let alone all of them.

We are getting a better handle on this, though, as we gain a better understanding of how different types of skin and cover relate to the underlying bones of the face. There have been a number of papers looking at these issues in living species in order to try and interpret dinosaurs and other extinct animals more accurately, and this does now give us a better feel for the presence of scaled skin versus roughened and thick keratin on various animals.

Beaks and lips?

A still greater puzzle, and one that is currently the subject of some heated debate, is the presence or absence of lips in dinosaurs (and especially, though not exclusively, the tyrannosaurs). Various theropods have been illustrated at times with either excessively toothy grins, giving them a crocodile-like appearance with their upper teeth on display, or a tight-lipped one where the teeth are completely concealed, as seen in modern lizards and snakes.

This may not strike you as especially important to work out (and many facts of dinosaur research have no direct and obvious consequences for other branches of science), but such questions are necessary to resolve in order to discover exactly what dinosaurs were like. Excitement from the media and the public might focus on new species, or bite power or speed, but animals are more than merely statistics, and completing the jigsaw of every detail of their appearance and their lives is what brings them to life metaphorically, and allows us to understand them properly scientifically.

Perhaps surprisingly, there are some strong arguments for both lips and lipless-ness, meaning, like many dinosaurian features, that it's not so much a guess as an unresolved issue. The main arguments come down to the ancestry of the dinosaurs versus what we think is more normal in terms of ecology. In terms of the former, the birds are not much use to infer anything about dinosaur lips, since all living species have a beak. The crocodiles are the next nearest living relatives of dinosaurs, so we should probably try to take our cues on lip and tooth anatomy from them if possible. They are very much lacking in the lips department, so the argument would seem to end there – dinosaurs did not have lips.*

The problem with this line of reasoning is that the living crocodilians are rather unusual, since they have had over a hundred million years to adapt to an aquatic (or at least semi-aquatic) and predatory lifestyle. Moving quickly in water relies on reducing resistance as much as possible – and a crocodile that lunged for its prey with any kind of lips would suffer huge resistance from any water sloshing into the open mouth and staying there. As a result, evolution has cut out their lips and so they may not be a good model for dinosaurs at all – ecology may have trumped ancestry and removed lips from crocodiles.

Although other living reptiles are a little further away from the dinosaurs in evolutionary terms, the presence of lips in almost all of them

---

* Or to be more exact, I should be saying 'extra oral tissue'. As with so many terms, this seems unnecessarily pedantic and creating a new term when English has a perfectly serviceable word, but lips are rather mammalian things that have muscles in them and are flexible and can move about. The equivalent tissue in frogs, lizards and (potentially) dinosaurs is pretty static and just sits there, hence the separate term to separate them.

(except those that have beaks with no teeth such as turtles) suggests that this might be the norm and thus most dinosaurs were lipped.

In the last couple of years, two new lines of evidence have been brought up to try and resolve this issue, though they support opposing positions. First off, it has been noted that animals with their teeth primarily exposed to the air lack enamel (or at least it is very thin). For most vertebrates, teeth are covered with a layer of super-tough enamel, but the tusks of elephants and various pigs, for example, lack this. The difference should help stop teeth drying out and cracking, so enamel loss would appear to go with an absence of lips leaving the teeth exposed to the air. Crocs do have enamel-covered teeth, but then they live in water, which should offset the issue of drying out.

If dinosaurs were lipless then their teeth should lack enamel, but as they do have a full enamel layer, this suggests the teeth were protected and, by extension, that lips were present. In contrast, the texture and structure of the jaws of a new tyrannosaur from Montana have been suggested to be very similar to those of crocodiles. This implies similarly large and thick scales like those that line the edge of the mouths of these animals and therefore that the dinosaur in question, at least, would have lacked lips.

One area where this kind of research would be especially weak is for the various herbivorous dinosaurs. Many, if not all, of these animals had a beak at the front of the jaw, and their main way or cropping or browsing vegetation came not from teeth, but from a large piece of keratin overlying the bones at the end of the snout. This beak was an early feature of the ornithischians and stayed with them for their entire evolutionary history. Many of them were rather large and extended the shape of the face with a saw-toothed margin quite unlike those we see in modern birds or tortoises.

In fact, beaks turn up in numerous reptile groups, not only the living chelonians and birds, but various other groups also had a beak. Some theropods were beaked animals, too, either combining a beak with teeth or possessing only a beak, including the bizarre *Limusaurus* – apparently a dedicated herbivore, but one of the ceratosaurs, a lineage otherwise composed of carnivores. *Limusaurus* is especially odd, as juveniles of the species had teeth and perhaps lacked a beak, and only later did they lose their teeth and stick simply to a beaked existence. Clearly in pretty much all of these groups, data from tyrannosaur lips (or absence thereof) will be of limited value to us in working out what their jaws looked like, beyond the presence of some tough material at the front.

The second major reason that the ornithischians would likely differ from large theropods in their lips is the nature of their teeth. In various groups the teeth lie in banks along the inside of the jaws and are set up to allow the animals to chew their food effectively, grinding up tough plants to aid digestion. Chewing without lips, or cheeks for that matter, would likely result in food spilling out of the side of the mouth. While dinosaurs didn't necessarily need to be the neatest of eaters, it does seem unlikely that these animals evolved such a complex food-processing apparatus and yet 'let' their mouthfuls spill out of the sides of the jaw with each bite.

While getting direct evidence to support this is clearly tricky, simple analogies with other herbivores suggests that lips would be likely among these animals, at the very least.

Feathers

All of these fleshy possibilities lie on top of the actual issues of the skin itself. We have dealt in some depth with the scales, but as we now know, many dinosaurs were blessed with feathers of various types and other feather-like filaments. Starting in the late 1990s, the first feathered dinosaurs began to be unearthed in China and we now have hundreds, if not thousands, of fossils representing dozens of species from China, Brazil, Germany, Canada and even North Korea that show traces of these different skin coverings.

All true feathers that we know of are in the lineage of the theropods. These comprise a number of tyrannosaurs and various more bird-like lineages, including the immediate ancestors and nearest relatives of birds such as *Velociraptor* and *Oviraptor*. Most of these animals possess feathers that are closer to those of bird chicks or birds like the kiwi – thin filaments that resemble long, thick and slightly stiff hairs, rather than those feathers seen on a chicken or eagle. But plenty of dinosaurs did have these more advanced forms of feathers, which include a shaft and various branches and sub-branches to makes up the 'sides' of the feather as seen in so many modern birds.

Various dinosaurs had other unusual types of feathers that are not known in modern species. Thick and near-solid quills appear along the neck and back of some herbivorous theropods, and these are thought to have offered some protection from the local carnivores in the way that the spines of hedgehogs and porcupines do. Other dinosaurs had ribbon-like feathers, which were solid sheets of keratin rather than being filaments or branched hollow fibres. Also, as seen in modern birds, some fossils show that juveniles had different feather types and sizes to those of adults, presumably moulting as they grew and shedding one layer for a different one as the demands of their lifestyle changed.

Of the many spectacular feathered fossils, we do have one issue that remains – the interplay between feathers and scales. Not only do we not know what the origins of feathers are (a highly modified form of scale, or a novel structure), but we don't understand if the two features are compatible.

At first glance, modern birds appear to have scales on their feet; feathers on most, or all, of the rest of the body; and nothing on the beak and occasionally bald patches on the head and neck. But in some birds it is clear that small and simple feathers *do* poke out between the scales on the feet and, to make life more awkward, bird scales are not like those of reptiles but are very odd feathers in their own right. At some point in their ancestry, birds lost all traces of the reptilian (and presumably dinosaurian) scales and went through a scale-less phase, before reinventing the wheel later on and converting feathers to scales. In short, we cannot say if scales and feathers were mutually exclusive.

A paper in 2017[*] that manipulated the regulator genes of scale growth in embryonic alligators and feather growth in developing chicks did manage to produce some very long and thin scales and some very odd feathers, which in some cases looked like those filaments preserved on some dinosaurs. The implications of this were

[*] Wu, P., Yan. J., Lai. Y-C., Ng, C.S., Li, A., Jiang, X., Elsey, R.M., Widelitz, R., Bajpai, R., Li, W.-H., et al. 2017. 'Multiple regulatory modules are required for scale-to-feather conversion'. *Mol Biol Evol.*, 35(2), pp. 417–430.

clear: feathers might be true scales that have been highly modified, although the evolution process by which this occurred may have been surprisingly simple. This is clearly very leading and informative and promises much more in the future. If it can be definitively shown that feathers truly are scales, this may finally show that the two were incompatible. In other words, when we find scales we could be confident that feathers were absent.

Currently, however, when palaeontologists find fossils of feathered dinosaurs that are mostly imperfectly preserved, it is far from clear if any scales were present. In cases where we have excellent impressions of fossil skin or even preserved skin, we do not know if there might have been some fine feathers poking between the scales that were simply not preserved in coarse-grained sandstones. Thus apparently exclusively scaly dinosaurs may have had feathers, and feathered dinosaurs might still have had scales.

This gets more complex (unfortunately) once we dive into the lineages with feathers. As far as we can tell, once a lineage had feathers they kept them. While many skeletons are preserved without feathers this is generally an issue of fossilisation rather than lack of a structure – when conditions are good enough to preserve feathers, we see them in every representative of feathered lineages. Exact degrees of coverings may come and go and there will be differences in the size and shape and even type of feathers, but feathers seem to be the norm for these groups.

Thus when various tyrannosaurs are known to have had feathers, and indeed been completely covered with them, the obvious conclusion is that all tyrannosaurs had feathers. Most species are known from fossils that have simply been recovered without them because they are not preserved, rather than were genuinely absent. There are only a couple of examples of even *bits* of feathers preserved in the rocks that *Tyrannosaurus* heralds from, for example, so it's hardly surprising that the tyrant king is yet to be known from a feathered specimen. But the logic holds up: it should be feathered, and probably well covered.

However, studies published in 2017 included bits of skin from a number of Mongolian and North American tyrannosaur species, including *Tyrannosaurus* itself, which showed the clear impressions of scales. This alone doesn't rule out feathers, but it certainly challenges the idea of a fully feathered *rex*, which had been the (justified) assumption for some time.

On a similar note, the various early theropod lineages are generally only known from environments that do not preserve fine structures like feathers (or even skin), and so the question remains: how far back do feathers go? Stuck with an absence-of-evidence problem, it's hard to say if feathers might have been rampant in the theropods, but simply not seen in most of them. Alternatively, feathers might have been a rather late 'invention' and thus proliferated once they had first appeared.

Feather–like things

So much for feathers in theropods, but there are other types of filamentous structures known in several ornithischian lineages as well. To date, none have been demonstrated to have been true feathers (i.e., sharing a single evolutionary origin with the feathers of theropods and birds), but the possibility remains. This feeds back into the question of dinosaur origins and the relationships of the great lineages of the Dinosauria. If the ornithischians really did have feathers, and they are the nearest relatives of the theropods, this implies the possibility of a single origin of feathers then inherited by (all?) their descendants. That would mean a great many more of these animals were likely to have possessed feathers of some kind, including all kinds of early theropods such as allosaurs and spinosaurs that we currently think of as being scaly.

On the other hand, if they are more distantly related, it becomes more likely that such features evolved multiple times and filaments might have been in some way 'easy' to evolve. The recent study on scale development mentioned above, for example, suggests that only a few minor changes to the genes that control scale formation might throw up filaments.*

---

* As an aside, the pterosaurs also had some kind of filament on their body, so in fact there is a third group to consider when it comes to the possible origin of feathers and filaments.

None of the ornithischians to date found with filaments show the complexity and diversity of form of the later theropods, but they do enjoy some variety of their own. Also, their appearance in different groups on the family tree is rather more scattergun than that of the theropods – two relatively early ornithischians have filaments, but then so too does the Cretaceous *Psittacosaurus* (an early member of the horned dinosaurs).

The presence of filaments in relatively early lineages gives rise to the possibility that these may have been universal in early dinosaurs and only lost in later groups, and *Psittacosaurus* suggests that filaments may have easily come back later. It's certainly not impossible that even the great *Triceratops* had some kind of shaggy filamentous coat across it's huge form, though we know that at least some of their close relatives were covered in scales.

Of great importance here is the possible relationship between these various filaments and feathers. Are feathers limited to the theropods alone, or are the ornithischian filaments one and the same thing but by different names? Might this extend even further: are the filaments on pterosaurs the same thing as well? As you might guess, we don't actually know (despite various claims of relationships between these different structures at one time or another and several researchers calling both ornithischian and pterosaur filaments 'feathers', which only muddies the waters further), but the possibilities are tantalising.

It is at least possible that feathers, or their ancestral forms, extend back to the Triassic and occurred many tens of millions of years before the first things we would recognise as birds ever appeared. The early years of dinosaurian evolution may have seen numerous different kinds of dinosaurs, and their relatives the pterosaurs, exhibiting new forms of fuzz and filaments, even if these were later lost in some and went on to produce true feathers in others.

One of the filamented ornithischians, the strange *Kulindadromeus* from Russia, is especially important in this discussion. Across multiple specimens (showing that it is no quirk of preservation or a genetic oddity, but a real feature), this little biped shows multiple types of filaments and scales on different parts of its body. Some even seem to show multiple small filaments attached *to* scales. If that is the case, then

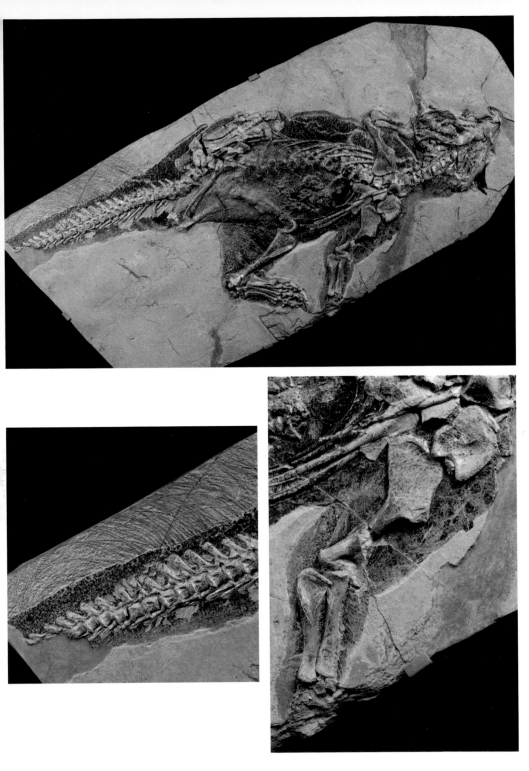

This spectacular specimen of *Psittacosaurus* has both scales from the skin and even feather-like filaments still intact and traces of the original patterns of the animal are still present.

A dinosaur feather preserved in amber that is over 100 million years old.

The bizarre *Yi* from China, this little glider combined feathers
with a flying-squirrel like wing membrane.

The superbly preserved armoured dinosaur *Boreaopelta*. Preserved colour traces suggest it had a dark top and was lighter underneath.

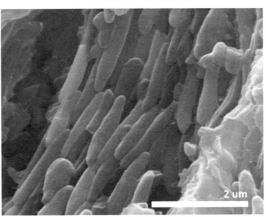

Fossil feathers, just like modern ones, preserve the same colour packets called melanosomes (these have been digitally coloured to make them stand out) that can be used to reveal the colours of dinosaur.

Known from only a pair of arms for decades, *Deinocheirus* turned out to be stranger than anyone had predicted with a large hump and enlarged lower jaw.

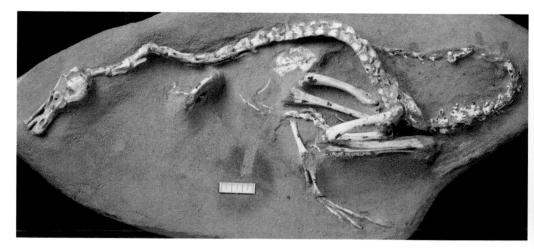

Halszkaraptor is the first, and so far only, dinosaur known that appears to have been a capable swimmer.

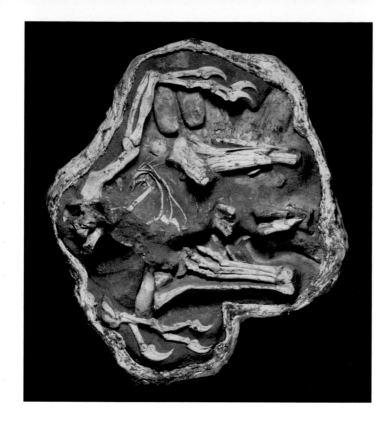

A mother *Oviraptorosaur* brooding her nest just like a modern bird.

This embryo of a *Titanosaur* preserved inside an egg shows the beautifully preserved skull of an animal that would grow to be more than 30 m long as an adult.

Dinosaurs can now be accurately reconstructed with missing tissues like eyes, nasal passages, brain and muscles, as in this *Allosaurus*.

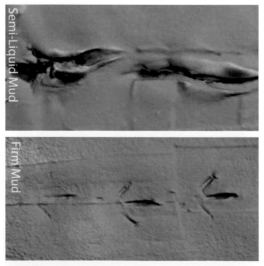

Semi-Liquid Mud

5 cm

Firm Mud

X-ray videos of footprints being formed by birds allow us to reconstruct the feet and movements of dinosaurs from 150 million years ago.

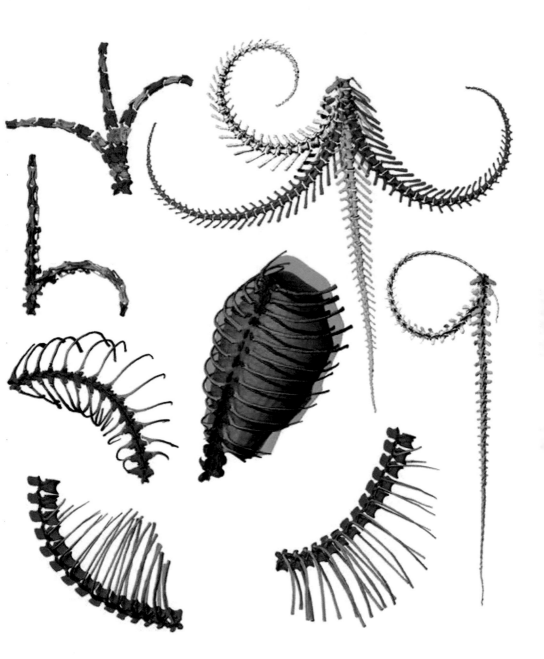

3D models of dinosaur bones allow us to reconstruct their ranges of
movement as in these parts of *Plateosaurus*. The neck (top left), tail
(top right and right), body and ribs (centre and left and bottom).

Even century old sites like Dinosaur National Monument are still revealing important new information about dinosaurs.

Femur of the most recent 'largest dinosaur ever found'. Just how big were these animals and what are we missing?

it opens up a different possible pathway for the origins of these fibres and perhaps feathers, too. It also opens up the possibility of things like the known tyrannosaur skin still having feathers coming from them – scales and feathers or other filaments might be perfectly compatible and not mutually exclusive.

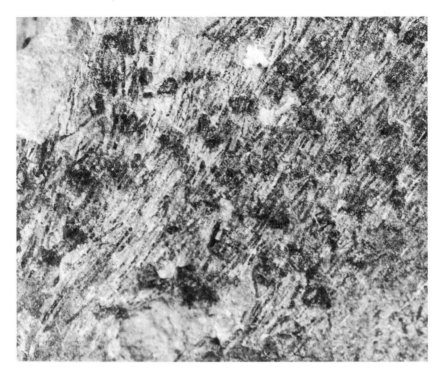

Some of the scales of the strange little ornithischian *Kulindadromeus*. These scales have filaments apparently coming off them. Photo by Pascal Godefroit.

In summary, the coverings of dinosaurs are extremely uncertain beyond the essentials of scales and filaments with some armour. We know many had scales of various kinds and many likely had lips or something similar, but not necessarily on the theropods. We know that multiple theropods had feathers, but not if the early ones were feathered. It's not clear if the tyrannosaurs began to abandon feathers later in their evolution, or if there were some complex mixtures of feathers and scales in various animals.

At least some ornithischians did have filaments and many more are known to have had scales, but how these might have been distributed among various groups and come or gone is a mystery, and their possible relationship to theropodan feathers is an enormously frustrating gap in our understanding. Happily, it is one that is attracting much research and is already being tackled by several research groups. The sauropodomorphs appear to be totally featherless, but given the possibilities of a plethora of early filaments in various early dinosaurs, it's not impossible that these animals did have them in the Triassic only to abandon them as soon as they started to increase in size, where insulation would be a positive disadvantage.

On that note, we suspect insulation was an early benefit of feathers or filaments, but it may well be that sexual selection and signalling of one form or another was a rather more important one. With that in mind, we can move now from the basic external appearance of the dinosaurs in terms of feathers and scales on to how these things would have looked once we include colours and patterns.

11

Appearance

LOOKING AT THE huge range of available data, we can build a pretty accurate dinosaur from most lineages and a large number of species. We have enough of the skeleton to predict the size and overall shape of an animal, the proportions of the limbs and tail, numbers of fingers and toes, the size of the eyes, the number of teeth and so on. We can also suggest an accurate posture, and degrees of motion of the various joints; to that we can add the internal organs and then the sheathing muscles and fat, and we can add on skin and any armour as well as scales and filaments of the appropriate type, size and distribution.

This would be an excellent representation of a living animal, but would be missing some key features – there would be a profound lack of detail on things such as the presence of soft tissue crests, and more obviously it would be monochromatic. What colour were dinosaurs?

This question has been a vexed one among palaeontologists for almost as long as there has been dinosaur research. Early answers of 'green, like reptiles' wear thin quickly and belie the huge diversity of colours seen in modern reptiles. Green is the stereotype and a common colour, but every colour of the rainbow and plenty more shades of blacks, greys, browns and white (and iridescence too) can be seen, as well as them appearing in all kinds of patterns and markings.

Even if you want to step away from other reptiles and stick to the archosaurs, the supposedly dull and monotone crocodilians have some striking stipes as juveniles and even a few of the caiman species as adults can be surprisingly brightly patterned. So there is no clear and obvious reason that all dinosaurs should be one colour, or for that colour to vary from green to brown (not to pick on *Jurassic Park* excessively, but has there ever been a more tediously coloured set of animals than as seen across those movies?).

Palaeontologists and artists have therefore been forced to base their interpretations on guesswork, and often it would be charitable to call it an informed guess at that. Sure, for all the bewildering colours of various reptiles (and birds), huge numbers tend to be browns, greys, greens and muted colours. Even those with vermillion, sky blue or acid yellow tend to have such bright colours in patches as signals, rather than covering most of the animal. However, it is the smaller animals that tend to be brighter colours and dinosaurs were nothing if not large on average, so perhaps the endless palate of greens and browns is not so inappropriate?

I don't think this is quite the case, as even some large animals can have at least patches of bright colours (the beaks of storks, the whole head of a king vulture, the pink flanks of hippos, the near orange of Sumatran rhinos), and a little more colour in some animals is surely to be expected. There is also a tendency to look at the largest mammals, in particular elephants, and see how uniform they are in colour (actually Asian elephants can have large patches of quite bright pink on them), but this overlooks the limited colour vision of most mammals so they have fewer colours to see. It also glosses over the fact that elephants are all but immune from predation and so camouflage is not really an issue. That may also be true of the largest sauropods, but plenty of the very large dinosaurs were clearly a target for theropods and some disruptive patterns would likely have been of benefit.

So the endless parade of grey sauropods and green theropods that covers almost every book of dinosaurs even into the present day should go, but what do we do instead? Until very recently, the one question a palaeontologist would find unanswerable would be, 'What colour were the dinosaurs?' There was simply nothing to add to: 'Probably more interesting than you have seen in books, but which were what colours and patterns, we don't know.' Now, however, we are starting to dig into this area and we do have some knowledge of the colours of animals from the Mesozoic.

This comes about because of tiny features of cells called melanosomes. These are essentially little packages of pigments that give colour to skin, scales, feathers and more. They are widespread in vertebrates (including humans) and so ancient colours are being described for more than just dinosaurs, though these are catching the headlines.

Melanosomes turn up regularly in various rock formations where conditions are suitable to preserve skin, feathers and the like, and so can be very common in some fossil localities but are absent everywhere else.

The key to their use for us is that each shape of melanosome seems to capture only a single colour type. It is as if you went to a paint store and each colour was in a uniquely shaped pot. You could find the empty pots and know what paints they used to contain. So while the fossil melanosomes have no colour now, we know what they should have held and from that we can work out the colours.

This is an incredible leap forwards in our knowledge and understanding, but it is also currently profoundly limited. Not all the colours of the rainbow are preserved in melanosomes; we can detect black, a form of black iridescence (as in starlings and magpies), tints of brown and red, and white. That's a great range and there are combinations of these we can also use (very dense sets of red are likely to be very bright, and red and black mixed are likely to be a dark brown), but we lack the ability to detect others.

This is because many colours are not formed from the melanosomes alone, but from their arrangement, too. These are termed 'structural colours' and come about because of the arrangement of melanosomes and not simply the pigments they contain. Since this delicate arrangement is hard to study even in living animals, it is difficult to have a good comparison with fossils and, on top of that, the patterns present now in fossils are not likely to be a good representation of the live condition – decay and fossilisation (especially compression in the beds where melanosomes generally preserve) will disrupt this.

So we are left with being able to make some general statements about some dinosaurs that are probably pretty good, but a bit vague. For example, the small compsognathid *Sinosauropteryx* from the Early Cretaceous of China has been described as 'ginger' with white stripes, and while there are lots of the red/brown melanosome on much of the animal, quite how much of this was an orange or red or brown, if it was muted or bright, and whether it varied across the animal, we don't know. That we have any information at all is amazing, but it is also frustratingly incomplete, since there's a big difference between an animal that is a real brick red or bright orange, and one that is a more earthy brown in tone and more likely to be using camouflage than trying to stand out.

One additional issue is that you need an extraordinarily powerful microscope to see melanosomes and one cannot fit a whole fossil into such a machine to look for them. That means you have to take a sample of the specimen to work on. This brings two issues with it: first of all you are actively damaging the specimen to access this information (though the samples are tiny and the data is worth it), but, second, it also means you can't sample everything. It would be too expensive and too destructive to take hundreds or thousands of samples from across a single specimen, and even a few dozen samples risks making some serious damage, however careful and experienced is the researcher.

That means that what we really have as data for various dinosaurs are lots of points distributed across the specimen. If, as with *Sinosauropteryx*, you keep finding alternate orange and white, it's reasonable to infer that the tail had stripes, but it's also quite possible you have missed some small patches of colour from other areas (perhaps a very thin black stripe that borders the thicker bands), or something like a colour change at the tip of the tail.

One other way of looking at colours can add to data from melanosomes and is not destructive, if yielding only some qualitative data. Intensive X-ray scanning of fossils can reveal patches with unusual chemical signatures. For example, copper is linked to certain red colours produced by birds (which is different to the ones in melanosomes) and would otherwise be extremely rare in a healthy animal. So traces of copper in some feathers (or potentially scales) could indicate that these were this colour, either wholly or in part.

Although not destructive, this technique can only be applied at a handful of sites in the world and means transporting specimens that are often very fragile and valuable long distances, when the results may be uncertain or non-existent. Thus, while it has its place, it is easy to see why this method has not been extensively taken up by palaeontologists.

At least some specimens do also show patterns of tone on them. Typically, areas that have no melanosomes at all or other pigment types and these can end up being white or at least pale. Some fossils clearly show a white or pale underside and a darker top. This type of patterning is called countershading and it is extremely common in animals of all kinds and acts as a form of camouflage.

The underside of an animal is almost inevitably in some form of shadow, as light will come from above, while the top remains well lit. However, with a pale underside and a darker top, these lighting effects are somewhat balanced out and the contrast is lost. By removing these differences in tone, the animal looks less like an obviously solid object and can blend into a background better and be generally harder to spot. This is true if you are a herbivore trying to avoid being eaten, or a carnivore who wants to avoid being spotted by their prey, and so it is little surprise that several animals already seem to show countershading.

Even so, we have little enough evidence as to what these animals' true colours were. We have worked on the patterns and colours of only half a dozen dinosaurs, and these are almost all small, feathered animals from a limited set of fossil localities. Even those we have worked on, our understanding is limited since the results are based on sampled points across a specimen rather than a complete set of coverage. Again, this is not to undermine the strides made but to point to the sparseness of the data, and in particular the gaps in our knowledge, which will make the data we do have so much more interesting and reveal so much more.

Only the beginning

A good example is the small and extremely feathered troodontid *Anchiornis*. Known from the Middle Jurassic of China, there are a very large number of specimens of this sitting in collections (sadly mostly in private hands and inaccessible to researchers), so many in fact that there might be more specimens of *Anchiornis* than of any other dinosaur. Many of them are complete, and preserved with feathers, though they are also typically squashed flat.

One that has been studied is described as being black with white patches and some red spots, and with a little red crest on the head. Huge numbers of different versions of this pattern have now been produced by various artists, and the internet is almost flooded with representations of *Anchiornis* that show these colours in this pattern.

They are not incorrect, but even allowing for the issues of taking colours based on 'points', this description is based on one animal, at one point in time, and may not be representative of the species as a whole at all.

It is normal for there to be some variation in the colours and patterns of living animals of the same species. Even if we are not very good at seeing it, there will be subtle differences between any two individuals and some are much more clear. For example, no two giraffes have the exact same pattern of blotches and reticulations, and there will be differences in tone of the main patterns as well as the degree of darkening of the horns and the tail. Some species are more variable than others, but variation is the norm, and some animals may have much more of one colour or another, or entirely lack one patch or one tone.

There are also plenty of major deviations of 'normal' out there, with animals that have reversed normal patterns or those that make far too little of colours, such as leucistic and albino animals that tend to be white, or melanistic ones that are mostly or entirely black. Such animals might be unusual, but are entirely normal in the sense that they appear through genetic or developmental quirks and may persist or even dominate some populations (such as black leopards).

Can we say that the *Anchiornis* we have studied is truly normal? Did a typical individual have much more red, or considerably less white, or was it more of a grey body than a black one? The odds favour that the animal we have sampled was fairly typical for its species, but there is no guarantee that we have not sampled something that was rather unusual for one reason or another and if so, the pictures we have painted of this animal might be giving a false impression of what they would normally look like.

What is much more likely to lead us into a false impression is the difference we might expect because of the age or sex of the animal. While hardly universal, a huge number of animals show major colour differences between males and females. Males are typically much brighter and/or have larger traits such as feathers, through which they advertise their social status and quality as a mate. There are no reasons to think such differences might not also be present in extinct lineages (and indeed we see at least some evidence for this in various dinosaur groups). As we will see, sexing dinosaurs is rather difficult, so is the one *Anchiornis* we have looked at a male or female? Either way, is this the same colouring as the opposite sex or different?

Even for a single individual, they would probably change their colours and patterns over time – juveniles would likely be more camouflaged than adults, males may be much brighter in the breeding season than outside of it, and males and females could potentially have a different plumage in summer and winter to maximise their ability to hide. All of these patterns of change are very common in various living groups (especially birds) and might well have been in operation in the past. We certainly do know that some feathered dinosaurs had different types of feathers between the juvenile and adult phase, so a change in colour and pattern might also be anticipated.

We can also expect evolution to have a say. Different individuals and populations will live in varying environments and at separate times. Over a few million years it is not hard to imagine that a species may become darker or lighter, add or remove some colours, become more or less patterned and so on. Denser forest in one place might select for dappled animals for camouflage there, but the grasses would lead to stripes elsewhere, or females might prefer red males in one region and look for males that colour and blue in another and so select for different colours. We see this again and again in living species, and to trace such differences it becomes critical to know the exact age and location of a given specimen. Sampling the colours of dozens of specimens from one species would be extremely interesting in its own right, but we would need the right contextual information to make the most of it.

In short, even if the colours of the *Anchiornis* or any other dinosaurs we have looked at have been identified given the uncertainty over tone, we can't begin to study changes over one lifetime, variation within a species or changes over generations, or what this might look like at the population level with males and females and all the varia-tion out there. However, these are things we will begin to discover soon. As more and more specimens are examined and sampled, we will start to build up profiles of whole sets of individuals from a single species and we can see whether or not these various differences are present, and if so which ones and what they mean.

Discovering that males are brightly coloured and females are not, for example, would immediately have implications for the reproductive behaviours of the different sexes and what this might mean for displays, pair bonding, rearing of offspring and others. Colour is fascinating to unveil in its own right, but the next level of our understanding is within our grasp.

To date, almost all the work on dinosaur colours and patterns has been on feathers, but melanosomes, or at least some of the chemical traces of the melanin pigments, do also preserve in scales and in keratinous sheaths of various kinds. Two ornithischians so far (the early ceratopsian *Psittacosaurus* and the ankylosaur *Boreaopelta*) have sufficient quality of preservation to show that they had a disruptive pattern to help hide them. It's only a matter of time before we delve further into scales and what they can tell us, but incredibly we might be grossly underestimating what they can do.

In early 2018, a team of researchers published a paper describing colour changes in living crocodiles. Chameleons might be the most famous of reptiles to alter their colour, but lots of snakes and lizards can change various tones and patterns on their skin, sometimes quite quickly. This appeared to be entirely absent in crocodiles and it is absent in birds, so it was never seriously considered for dinosaurs. However, if some crocodiles can do this, then dinosaurs at least had the potential capacity to evolve colour changes, or may have inherited it from their ancestors. The possibilities are extraordinarily intriguing.

The function of colour

Aside from disruptive patterns and sexual and dominance signals, there are other possible patterns that could be adopted by various dinosaurs. Melanosomes are known to help provide support and strength to some features and it's possible that the reason some early gliding dinosaurs and birds had predominantly black feathers was because these were stronger and more resistant to damage. We might similarly expect the soles of feet, claws, or the cloacal opening to have been black, as this would help them to resist some wear and tear in daily life.

Dazzle camouflage works by having bright and contrasting bold colours that can make it difficult to estimate speed and direction of travel. This has commonly been applied to warships in the First and Second World Wars, but it seems to be mostly absent in animals. Zebra stripes are often supposed to operate in this manner, but studies have shown that predators are no less successful hunting zebras than other species. There is some evidence that zebra stripes can help ward off biting flies, so such a pattern could still be of value to some dinosaurs. There were plenty of blood-sucking animals around at the time and so stark stripes might well have had some unusual benefits.

Bright and contrasting colours and patterns can be used as a warning signal. Instead of trying to hide, you can actively advertise yourself as something to avoid. We see this mostly in venomous and poisonous animals, but there's no reason to think that especially well-armed dinosaurs like ceratopsians and ankylosaurs might not have been very colourful. When such features do take hold, other species may come to mimic them. If, as a species, you are also well capable of defending yourself from attack, looking like something else that also does can help you both. A predator need only encounter one of the two to learn that it is not a good option and from then on would avoid any creature like you.

The more species that carry the same warning and the same deterrent, the faster and easier it will be for carnivores to learn. This is called Mullerian mimicry, and can drive the evolution of similar signals among multiple species, even if they are unrelated. It is certainly possible that where multiple species lived together – and in the case of the ceratopsians in particular, there were many horned species that are found in the same fossil beds – these may have acquired very similar colours, patterns and even stereotyped displays. Again, such patterns are likely to change during growth. A baby *Triceratops* lacking the horns of an adult would be much more interested in hiding than in false advertising that might give it away.

A second common form of mimicry is that of Batesian mimicry. Here a species evolves to look like a dangerous one, even when it lacks their punch. A hornless ceratopsian may evolve to have the displays of an armed one, playing off the fact that predators will avoid them based on their knowledge of the others. This again is found in many living species, but we only see it when those mimics are relatively rare. If they are common, then predators will encounter them more often than the dangerous ones and soon learn that these signals are not backed up by a real threat.

As such, we may predict that we will find them only when a relatively defenceless animal lived alongside a more threatening one, and was also less common. As yet, we have not seen any of these appear, but they are prevalent enough and well enough understood in extant groups that they are something that could well have evolved, and with enough data we may yet be able to demonstrate their existence.

Other parts of various animals have colour and shape, too, most notably the eyes. The irises of reptiles and bird are very diverse in terms of their colours and patterns. Many have remarkable patterns and tone, with everything from near jet black to the brightest reds, greens, yellows and blues. This gives us no immediate limiting factors on dinosaur eye colours (and even looking at crocodilians on the reptile side, there are also some bright colours there), though we might expect those who went for more muted colours and disruptive patterns to have had relatively dull colours to avoid being easier to spot, while some might have been selected for brighter colours in order to advertise.

There is also a great deal of variety in eye pupil shape in various animals. Pupil size is correlated with eyeball size and we can work this out from the size of the orbit in the skull. Large eyes can give you very acute vision (good at seeing small things, or that are a long distance away), and work well at night or in other low-light conditions. While the most familiar shape to us is a round pupil, since we have those and so do many common domestic species, they can be vertical or horizontal slits, and goats and a few others have especially odd rectangular pupils.

We are beginning to understand how different pupil shapes can evolve in response to differing situations (in mammals) at least, and from this we might expect round pupils in animals that are tall and/ or diurnal; vertical pupils in those that are nocturnal and/or are not tall; and horizontal pupils in those that are predated upon. There are many exceptions to these rules, and there are some potentially conflicting signals (tall nocturnal species, for example), but they are common enough that they might well apply to dinosaurs too, and at least offer some strong hints.

A few extremely well-preserved fossils from the Lagerstaetten beds show the eyes, though they do tend to be rather flattened. Detailed analysis has revealed even the cellular structures of some fossil birds from these beds, so it is not impossible that we will be able to work out the pupil shapes of some dinosaurs in the future.

Sight isn't the only sense that might have led to some additional changes or features popping up on dinosaurs to add to their appearance. Various birds have thick and bristle-like feathers around the margins of their mouths to act as sensors of one kind or another, which act as whiskers do on mammals. Various feathered dinosaurs might well have employed something similar, especially those that foraged at night, or gliders that tried to snap at things while on the wing.

Similarly, owls have facial discs of feathers to help channel sounds and improve their hearing. While this is a fairly specialised feature, it could have been present on some dinosaurs and at least one alvarezsaur (the little anteaters) has owl-like asymmetric ears, with the two sides being on different positions on the head. Given that these were small, feathered predators with large eyes, and likely could see well at night, there is something to be said for the possibility of them being very owl-like in appearance.

Finally, many birds have small stiff feathers around the eyes that act as eyelashes, helping to protect the eye from dust and grit. These are prevalent among larger birds with large eyes and those that might find grooming more difficult, and it's quite possible these were also common among various feathered theropods, but we'd need some exceptionally preserved specimens to work from and they might easily be confused with other feathers on the head.

Feathers, as already mentioned, could also shape the appearance of an animal further. We already see evidence for erectile fans of feathers on the tails of oviraptorosaurs, making them like mini-peacocks,* and

---

* Though for the record, the train of the peacock is not made of tail feathers, but is from the animal's back, and then supported by the tail, which is erected behind it to hold up the flashy long iridescent plumes with the eye spots.

several gliding dromaeosaurs have long streamers on the tail, as do many modern birds. Different feather types, especially such enlarged ones, would make a clear difference to the looks of a dinosaur, but other more subtle and important ones might be present. *Anchiornis* and several other feathered theropods have been suggested to have had tufts or crests on the tops of their heads, as do modern cardinals and cockatoos.

Certainly they appear to have such a thing based on their fossils, but living birds that lack such crests can also produce something remarkably similar once dead and dishevelled and then crushed flat. Experiments show that these features can be reproduced remarkably closely from animals that have perfectly smooth and rounded heads of feathers when alive, so it's rather hard to tell.

We are still not quite done with soft tissue bits on the heads of dinosaurs, as is well illustrated with birds such as modern turkeys and chickens. While many dinosaurs exhibit a wonderful array of horns, frills, crests and other exaggerated structures on their heads, these are bony in nature and it is fairly easy to have a good idea of what they were like when the animal was alive. But turkeys and chickens (as well as various other birds, including a number of vultures, cassowaries, ducks, storks, and perhaps unsurprisingly, wattlebirds) have their wattles and combs.

These are fleshy extensions (sometimes both large and pendulous) that as far as we can tell have no relationship to the underlying bones of the skull, or vice versa. In short, a turkey skull in gross terms is indistinguishable from that of other birds that have no such features. Thus, while plenty of dinosaurs like the ceratopsians do have bony crests, this doesn't mean that their apparently simple-headed relatives had smooth crania.

Indeed, a recent discovery of the large hadrosaur *Edmontosaurus* from Canada shows precisely this – a form of fleshy cockscomb on the top of the head composed entirely of skin. An animal that was one of the bony crestless groups of hadrosaurs in fact bore a crest. This opens up enormous possibilities for both these 'crestless' dinosaurs and many other lineages. There are plenty of great examples of birds with wonderfully colourful and elaborate fleshy crests, or even bald patches of skin that are brightly coloured. There are the wattles of various pheasants and turkeys (and in particular the beautiful tragopans), inflatable sacs as in frigate birds (or elephant seals for that matter) could also be present, and birds from cassowaries to birds of paradise via various waders, storks and vultures take this route into advertising.

New fossils continue to be found that reveal details of the appearance of dinosaurs. Here, the tail of *Edmontosaurus* shows the ridge of little serrated scales along the top. Photo by John Scanella.

With so many avian dinosaurs evolving such features (and at least one crocodilian, too – the enlarged snout of the male gharials), it would be a surprise if at least a few of the vast variety of dinosaurs hadn't gone down this route as well. Again though, we lack any direct evidence for anything more elaborate than the small hummock on an *Edmontosaurus*, but even more of an issue remains: if we did find them, would we even know?

An exceptional mummy or other specimen preserving skin could turn up with a big flap of skin around the neck or the top of the head, but it would be hard if not impossible to work out or show if a given structure was passive, or if it was capable of changing shape or inflating, or being moved by some muscles. Short of a new, and remarkable, deluge of dinosaur mummies being discovered, this remains a frustrating gap in our understanding, not least when we are only beginning to understand the functions and importance of the bony crests.

At least some dinosaurs were brightly coloured and patterned, and others were more muted and had evolved patterns that would keep them hidden rather than standing out. Such a mixture may not be much of a surprise, but it stands in contrast to the near endless parade of grey-green monochrome monsters that still adorn far too many illustrations and incarnations of the dinosaurs. As more and more data accumulates on colour and patterns of these animals, and equally importantly, as biologists get a better handle on the evolutionary pressures that drive colours, patterns and signals in living animals, we will begin to tease out how we can best predict and understand the colourful and elaborate world of dinosaur communication.

12

Reproduction

REPRODUCTION IS FUNDAMENTAL to all life and over their 130-something million years on Earth, dinosaurs went through many, many generations and the production of innumerable little dinosaurs. Something so universal has inevitably attracted considerable research interest from palaeontologists, but this is an area that combines various aspects of dinosaur biology that are very hard to study, most obviously their behaviour and soft tissue anatomy. We didn't even have any positively identified dinosaur eggs before the 1920s, so it took a very considerable amount of time after dinosaurs had been discovered for us to start piecing together how they might have reproduced. As a result, this is a field of study with some of the most puzzling and uncertain areas, which is something of a pain given just how essential reproduction is as a part of the biology of any organism.

Males and females

To begin at the beginning, if a mummy dinosaur and a daddy dinosaur love each other very much, they will get together and produce baby dinosaurs. But can we even tell mums and dads apart? The answer is actually 'Yes', although a better answer is, 'Yes, well sometimes, depending ...' We can be confident that dinosaurs did have males and females, since all birds and crocodiles have this form of reproduction with fully separate sexes.

Various lizards have female-only species, or even populations of species that lay fertile eggs, and some others are capable of this kind of reproduction when no males are around (including Komodo dragons). However, this is unknown in the archosaurs and so dinosaurs presumably had two fundamental sexes. Similarly, some lizards and snakes give birth to live young, but no crocs or birds do, so it's safe to assume that dinosaurs were all egg layers. We have fossils of various Mesozoic reptiles that did give birth to live young with the skeletons of their babies inside them, but this has never been seen for any dinosaur, which supports the idea that they were egg layers.

During the breeding seasons, a sexually mature female reptile or bird will lay down an unusual form of very porous bone on part of their skeleton called medullary bone. When laying eggs, they will need rapid access to a large supply of calcium to make the shell, and medullary bone serves this function. Dinosaurs did exactly the same thing and so we can take tiny samples of their fossil bones and look for the key features of medullary bone and, if present, then we clearly have a female dinosaur.

So far so good, but the absence of medullary bone doesn't indicate that the fossil is a male. It could be a female outside of the breeding season, or one who has already laid her eggs, or was too sick to breed that year, or hadn't reached sexual maturity yet. So even finding a whole bunch of skeletons together and being able to show that a good number of them were female, and the breeding season was in full swing, still doesn't make the rest males. We're left with a test that, if it proves positive, denotes a female, but that leaves a plethora of possibilities if it is negative.

One might anticipate that there would be some clear indications of males and females from the bones, since so many animals including all kinds of reptiles and birds have males and females that differ dramatically in their sizes and shapes – a phenomenon called sexual dimorphism (literally, two shapes). Here things get decidedly sticky, since first of all most dinosaur species are known from only one specimen, which doesn't allow for much in the way of telling apart the sexes, but even where there are good sample sizes, there have been no good demonstrable examples of dimorphism despite various claims to the contrary.[*]

My own work with colleagues[†] suggests that dimorphism is usually

---

[*] *Tyrannosaurus* inevitably looms large here and at various times it has been suggested that females were larger and heavier than males, or were generally just bigger built, but this doesn't hold up to any rigorous statistical analysis.
[†] One of whom is my friend Jordan Mallon who, on kindly reading and commenting on the first draft of this book, amended the text at this point to state that he is 'my most esteemed and really cool colleague'.

too subtle to pick up with the kinds of sample sizes we have for even the best represented dinosaur species, and we would need many more skeletons in general – and for there to be good numbers of both males and females – to be able to start reliably telling the sexes apart.

It is still notable, however, that the most distinctive features of many dinosaurs (horns, frills, crests, plates and so on) are apparently present in all members of a species. We might anticipate that, following many groups such as deer or birds of paradise, males would possess some display feature or other that females lack, but this appears not to be the case. For example, we have yet to find an adult ceratopsian without a frill and while there are hadrosaur fossils that lack head crests, these are clearly different species to those that have them.

This seems odd, but while sexual dimorphism is common in living groups, it is far from universal, and the presence of some kind of signalling device on both males and females turns up often, too. Typically, males end up with some kind of ornament or display behaviour (or are simply bigger), because females are investing heavily in laying eggs, looking after the young and so on, and are therefore on the lookout for the best possible father (genetically at least) for their young to maximise their reproductive success.

Males that are freed from any investment can put all their energies into showing off in various ways and competing with each other to try and secure the chance to mate with more females, and thus improve their success. But conditions where the work of both parents is essential to raise any offspring puts males in the same boat as females – they don't want to be lumbered with a poor-quality female, meaning their babies will have a worse chance of survival and leading to males looking for a signal that a prospective mate is a good choice, likewise. This means both sexes are signalling to the other and both end up evolving similar ornaments.

This interpretation is well supported by various lines of evidence but it does still leave some major questions unanswered. Even allowing for the large numbers of species known by only one or two fossils, the huge variety of dinosaurs and the places they inhabited means that it would be a surprise if no dinosaurs showed strong sexual dimorphism, but which ones did we don't know. More importantly, what evolutionary selective pressures led to so many species apparently taking the route of mutual investment in their offspring? It's not uncommon among modern archosaurs, but for it to be so common (even just among the species where we have good samples) seems unusual and this is an area that requires much more research.

Most of that, sadly, can't come without much better datasets for key species, but it is an area of increasing interest and also one where more and more work is being done by biologists on living species, which will provide a much better framework for understanding how these reproductive strategies evolve.

Cue Barry White

So we may not have sorted out our males from our females (though we know they exist), but next they will need to pair up. Working out what courtship behaviours dinosaurs may have employed will lead to enormous speculation. Clearly those with elaborate frills or feathers would have flashed and displayed them in some way, and those well suited to interesting calls, such as hadrosaurs, likely made the right noises, but what that involved exactly is impossible to know. Living animals go through all kinds of incredible and bizarre routines and it would be strange indeed if no dinosaurs did anything other than just wander up to each other (though some probably did little more than this).

The armoured ankylosaurs, for example, have bizarrely convoluted nasal passages running through their skulls and these are speculated to have been connected to little inflatable sacs on the nose, which could have blown out like the throat pouches of frogs and frigate birds. Certainly there is something going on there that is not connected to any other obvious part of their biology and could easily come down to some sexual or social signal. Similarly, it's been recently shown that, quite incredibly, the bones of some chameleons fluoresce under the skin. These species have little frills rather like ceratopsians and, while a stretch, it's perfectly possible that some dinosaurs, with their excellent vision, did something similar to increase their visibility to members of the opposite sex.

There are all manner of other possibilities that are as impossible to discount as they are to confirm, with different scents, calls, gift giving, dances, elaborate buildings of anything from nests to towers of stones that dinosaurs may have carried out, and there would have been differing degrees of male and female involvement and potentially complex interactions between them.

In terms of interactions, some of the few things we do know about are those that leave clear traces on the bones. *Triceratops* have long been suggested to battle each other using their horns and there is excellent evidence to support this idea, as these animals show various scratches and punctures on their skulls from exactly the kinds of places where you'd expect their horns to impact on each other. The issue here is not what was *Triceratops* doing, but what about the other dozens of ceratopsians that don't show such patterns. Were they striking at other parts of the body, or not fighting at all and simply showing off their horns?

Similarly, many large theropods, but in particular tyrannosaurs (about half of adult tyrannosaurs show them), exhibit various healed bites across their skulls. Indeed, the rather unusual textures on tyrannosaur skulls suggest that they had large patches of armoured skin, and injuries on the bones would be even more frequent were it not for this protection.

While various carnivorous species surely got into a scrap from time to time, mostly these incidents did not turn into serious fights as one animal would back down; the exception being when this was a combat between members of the same species. The fact that injuries are often appearing on the face implies that these animals were either head to head or side by side and this indicates some kind of ritualistic behaviour to the interactions.

The issue that remains to be resolved here is quite what is leading to this and who is doing what to whom. First off, such engagements may not necessarily be the direct result of sexual interactions, but may be part of more general fights over food, territory or other resources; and second, these may be fights between males on females, or between members of the same sex battling each other. Resolving this is likely to rely again on us being able to correctly identify males and females (imagine if it showed that all of the injured animals were females, for example), so while this remains frustratingly unknown, it is also perhaps solvable once some of the problems of determining sex have been worked out.

Occasionally, new discoveries can come out of nowhere and open up new possibilities that had not previously been considered as possible, because we'd never find the right fossils we might need to demonstrate them. One such example is the recently described court-ship 'scrapes' that have been identified from Cretaceous rocks in Colorado. These consist of numerous pairs of furrow-like excavations that, based on their size and shape, were made by large theropods.

They are an incredibly good match for similar traces made by vari-ous living seabirds, including some plovers, which are formed when males and females do little ritualistic dances with each other and scrape away at the ground while doing so. It appears that at least some theropods did something very similar, and we can only hope that more sites like this will be found (or perhaps they have, but not been recognised for what they may show) and will shed more light on the possible courtship behaviours of the dinosaurs.

With the dinosaurs paired up, they then needed to mate. In many reptiles and birds this is a fairly simple pushing together of the respec-tive animal's cloaca, allowing the male's sperm to reach the female. For many dinosaurs, especially smaller species, this was surely sufficient for them to mate successfully, but there are immediately some obvious issues for other groups. How on earth are you supposed to get two ungainly and very prickly ankylosaurs together, or some of the giant multi-ton bipedal theropods, or the biggest of the sauropods?

Aside from the manifest issues of weight and a general lack of agil-ity that comes with being very heavy or walking on two legs, even before another giant climbs on top of you, to consummate the liaison those openings need to meet up. That is going to be very difficult if not impossible given the large and thick tails, broad pelvises, and other issues like spikes and sail backs and other such anatomical quirks.

Various postures and positions have been hypothesised for different dinosaur species to come together, but they remain largely untestable. We do have good ideas of the range of motion of a given set of joints (like the base of the tail and the legs at the hips) and can work out whether or not a given coupling would allow the animals to balance or tip over, but it's hard to do more than that. Without a set of tracks that might show exactly which foot went where (and leaving aside the obvious question of whether we'd even recognise a set of prints of a mating event for what it was), it's hard to say much about these ideas beyond them generally looking reasonable.

We know dinosaurs did mate, so however awkward it might have been, it clearly happened in every species many times. The obvious evolutionary solution to the gaps and at least some of the less likely-looking positions would be an intromittent organ, that is, some kind of phallus.* Such a feature has evolved many times independently in birds and reptiles and these can be proportionally very long,† and as a result any such feature would bridge the gap between a male and female that could not otherwise manage to copulate successfully.

All of these details are, of course, a mystery and it's hard even to work out which species may have ended up requiring an intromittent organ, but as we've seen in Chapter 7 on anatomy, it is at least possible that an exceptionally preserved animal may yet be found with sufficient soft tissues remaining to show one.

Nests and eggs

Eggs would then have been laid, but there is considerable uncertainty about how many of them there would be. It would be nice to know how many eggs a dinosaur laid, but matching eggs to skeletons is generally very difficult and without an embryo inside or an adult on the nest – and so working out whose egg is whose – it is near impossible. Unfortunately, even when we do have a nest brimming with a huge numbers of eggs and know which species laid them, it may not tell us as much as we'd like. The clutches of birds can vary enormously even for a single individual, but worse, there are birds that are communal nesters with multiple animals laying in a single nest.

Ostriches, for example, will lay a few eggs in a number of different nests, each of which are built and presided over by a male, so while he may be brooding a couple of dozen eggs, it may represent (part of) the

---

* But not a penis, since this is a mammalian trait and the various equivalents in reptiles and birds evolved independently and multiple times. So the name is kept separate to indicate that difference rather than due to any form of prudishness, though you will see the term penis used in various sources for these organs.
† I cannot but recommend you look online at the very long and explosively inflatable, corkscrew-shaped phalluses of ducks. It will be *very* educational.

reproductive output of numerous females. Therefore, working out what kind of reproductive effort is represented by a nest of dinosaur eggs is very difficult and is the kind of problem we may not be able to solve for more than a handful of species in the long term.

However, we can say rather more about the construction of the nests themselves and the incubation of the eggs. Most dinosaur nests seem to have been relatively simple affairs and were generally more like those of modern crocodiles than modern birds and would have been a hole in the ground or a pile of sticks with some eggs dumped in them, rather than some carefully arranged and intertwined affair.

Apart from a few of the more bird-like dinosaurs, they would have been unable to brood their eggs. Their size and proportions and lack of feathers would have made it impossible for a *Stegosaurus* or a *Brachiosaurus*, for example, to sit on a nest and keep the eggs warm; so a hole in the ground with some rotting vegetation for heat would have done the trick, and there is evidence for this in some hadrosaurs at least. For many of the small feathered theropods, though, we do see more carefully arranged and structured nests, with the eggs laid in a ring and partly above ground, and with space for an adult to sit in the middle. In fact, there are several specimens of various species with an adult theropod sat in the middle, arms spread out where in life their feathers would have covered the eggs.

Nest of eggs of a giant oviraptorosaur from northern China. The eggs are laid in multiple layers in a ring and the animal likely sat in the middle. Photo by the author.

At least one group of dinosaurs took a fairly novel approach to the issue of incubating their next generation. In the group of sauropods called titanosaurs, their eggs are usually found at sites with huge numbers of eggs where large numbers of animals must have been laying in the same location year after year. There are places where one can walk on almost nothing but dinosaur eggs in any direction for hundreds of metres. Titanosaurs have unusual compositions of their eggshells and, coupled with the locations of these massive beds of eggs, there is good reason to think that they were using geothermal energy from local sources of volcanic heat near the surface to incubate them. There are some modern birds that do the same thing on volcanic slopes, so there is a good precedent for such behaviour. This is another great example of something utterly unexpected, as despite knowing that some birds do this, it would not necessarily have evolved in dinosaurs and we may not have been able to find evidence for it, so it shows just what can appear unexpectedly in the fossil record.

Still, knowledge of what a handful of groups were doing doesn't tell us much about the rest. There are dozens of dinosaurs for which we have no eggs (or at least we have eggs where we have no idea to whom they belong). As a result, it is hard to say very much about what most dinosaurs were doing and what kinds of eggs they laid, how large they were, what their nests looked like, or how the eggs were incubated. That's a long list of very basic details for the vast majority of dinosaurs and represents a very sizeable gap in our understanding of their reproductive biology.

Our data on development is otherwise rather limited at the moment, but a recent study using microscopic lines laid down daily in teeth shows that at least some dinosaurs were in their eggs for anything from three to six months, an extremely long time. If this was the case for many species, it has important implications for dinosaur reproduction. Since parents guarding nests is common in living birds and crocodiles (and might be expected to be a shared duty in dinosaurs too, based on the evidence for mutual sexual signals), this means both males and females might have been hanging around the nest for months at a time and would pose challenges for at least some species to find enough to eat.

Might parents have taken it in turns to look after nests or even forage for each other? It's obviously very hard to say, but this is an area that is going to be explored greatly in the coming years and this kind of investment required for so long would be most important to understand.

Raising the kids

Certainly some dinosaurs definitively did look after their nests and even the young once they had hatched. Critical hadrosaur nesting sites show young in the nest far beyond the size they would have been when they hatched. That can only mean that one or both of the parents were bringing them food and that would have happened only if the parents knew exactly which offspring were theirs and the babies had stayed at the nest the whole time.

Various other dinosaurs occasionally preserve groups with a single adult and many young together, or with youngsters among groups of adults, further suggesting that for these there was some extended parental care after hatching. Others probably abandoned their eggs though (like the volcano-brooding titanosaurs), and we have various groups of young dinosaurs preserved together without the presence of adults, indicating that these can only have had parental care for so long.

What controlled this variation is not understood at all. Doubtless various degrees of parental investment, the nature of local predators and risks associated with them – adults can protect juveniles if they look after them, but then juveniles can attract predators putting adults at more risk – and how many offspring there were, would have been an influence. Animals such as hadrosaurs that may have nested in colonies (at least some did) could have flooded the landscape with juveniles, making easy pickings for predators but ensuring that at least some survived. There is a myriad of tactics available to animals and many will change their approach according to local conditions, so working out what may be the case based on some fossils may not even be true for most members of that species, let alone any others.

These potential different degrees of parental care have very important implications for the ecology of species and entire ecosystems. A dinosaur that has relatively few eggs and puts a lot of effort into rearing and protecting those that hatch will have a high survival rate, and so the population will typically consist of adults and a few juveniles. However, those that go for huge numbers of eggs and limited or no parental care will have populations of huge numbers of juveniles, which would rapidly be reduced by predators before the next wave the following year. These are two potential extremes and there are all manner of intermediates, but having an idea of where some species (especially dominant ones like the giant sauropods of the Late Jurassic) lie on this continuum is critical to understanding their reproductive strategies, and therefore what happens at bigger levels.

If *Apatosaurus*, *Brachiosaurus* and *Diplodocus* populations are typically 90 per cent juveniles, that makes big 10–20-ton animals relatively rare in the landscape and most of the biomass would be in juveniles of a ton or less. That would mean predators had large numbers of relatively small and vulnerable animals to approach and adults might be more or less immune to predation, while other, smaller herbivores would be in competition with numerous baby sauropods that were a similar size to them. If the reverse was true, though, with juveniles well protected by adults and there being fewer of them, but many large adults, predators would have to target bigger and better defended animals and the smaller herbivores might have rather more breathing room without a panoply of competitors. Either way, this would have profound impacts on the evolution and ecology of other species and so the reproductive biology of one group can potentially sway and alter the biology and interactions of many other species.

Unfortunately, baby dinosaurs are very rare in the fossil record for a number of reasons. First off, they tend to be eaten and their bones destroyed by predators, next because their bones are less well suited to being preserved as fossils, and finally because they are harder to find now being small. That means that regardless of these different strategies, the fossil record remains frustratingly sparse and it is diffi-cult even to begin assessing what degree of care, or what degree of population structure, might have existed for various dinosaur faunas. Only in locations where there is exceptional preservation do we find juveniles in any kinds of numbers and then they seem to be common, which supports the general idea of there being many juveniles. Whether or not this is due to limited care or very high numbers of eggs, or some other factor, isn't known.

As juveniles grew, they would also mature and themselves become able to reproduce. In mammals, getting close to final adult size and becoming sexually mature generally happen at about the same time, and birds are broadly similar. Crocodiles and many other reptiles, however, are capable of reproduction well before they get to full size and are still growing and show other signs of immaturity, such as unfused bones in various parts of the body.

Many dinosaurs were more like crocodilians in this regard, for example, we see small animals with unfused bones possessing medullary bone, showing that they were sexually mature even if they were not otherwise physically mature. This again supports the idea that many of these animals died at a young age, as reproducing early is a common strategy for those species that don't typically get very old, though as ever there are many aspects to this.

As with so many issues covered in this book, the main problem here is a lack of overlapping data. We have some excellent information on the nests and numbers of eggs laid for some species, data on the degree of parental care in others, onset of maturity in some and then information on population structure for still more. However, none of these datasets or the information on them is especially good given all the caveats, and then we do not have all of these different aspects covered for a single species, let alone many.

So while various analyses and bits of data point towards many dinosaurs having numerous offspring with some parental care and populations with numerous, vulnerable juveniles, it's difficult to support this very strongly, and it clearly doesn't apply to all species and may not even apply to that many. There are clearly exceptions and some of these (like the titanosaurs, with apparently no care) are especially important in some ecosystems that could have knock-on effects for so many other groups.

On the upside at least, this is an area that should improve as new fossils are found and give us more insights into these areas. Although it may be some time before there is much resolution across dinosaurs as a whole, our knowledge of the reproductive biology of many species is growing rapidly and this is an area that will grow substantively in the coming years.

13

Behaviour

O NE REPEATED THEME of this book is the apparently contradictory
situation that palaeontologists know both a lot more and a lot
less about dinosaurs than most people realise. Nowhere is this more
true than in the field of dinosaur behaviour. The public seem to be
under the impression that we have an excellent sense of how dinosaurs acted, with portrayals in the media (both fiction and documentaries) filled with commentary about this species or that being aggres-
sive, or social, or cautious, and elaborate ambushes being laid by
predators to single out individuals from a structured defensive arrangement of herbivores.

Most of this is pure fiction, and while often quite reasonable (surely
many theropods were capable of setting an ambush simply by hiding
close to water sources), there's no direct evidence for it and it can't
easily be inferred. On the flipside, we have worked out some incredible detail of dinosaur behaviours in terms of things such as feeding
methods, resting postures, nest building and even courtship rituals,
which seem impossible to have determined from the fossil record.

Thus, public perception of dinosaurs is almost the inverse of reality
and much of this chapter will be devoted to aspects of their behaviour
that you might think were solved long ago, such is the number of
times they have been repeated in various formats. If you are going to
deal with behaviour, the first and most obvious question is: exactly
how intelligent were dinosaurs?

Bird brained?

Here comes the first issue, since it is very hard to work out how smart living creatures are. Even with animals like primates, which are very similar to ourselves and that we have extensively studied, there is great disagreement over their capacity to carry out various tasks and grasp concepts. Move on to dolphins or parrots and things get much harder – they are clearly clever, but how smart and how they compare to each other is nearly impossible to determine, so doing the same with animals that have been dead 65 million years is a huge hurdle.

We can, though, try and work out some fundamentals. One of these is the oft-quoted (and really not that accurate, but useful) encephalisation quotient or EQ. This is essentially a measure of the size of the brain of an animal relative to its body size. Larger animals tend to have larger brains because they have more body to control, so there's a factor that needs to be controlled for, but one can look for animals that sit well above or well below the expected values and get a feel for how smart they are.

Humans sit well above the expected number for an animal of their size, as do primates generally. Most birds have an unusual brain structure, which means they can pack in a lot more cells into the given space, so they are typically rather smarter than their EQ would suggest, and that also means there's the possibility that some dinosaurs did this, too. It is unlikely, given that this appears to be an evolutionary innovation that appeared well after the birds split from their ancestors, but it can't be overlooked.

Based on the EQ, we would generally find the sauropods on the lower side of things (no surprises there) and some of the smaller theropods in the upper ranges (also no surprises). Some of the very-nearly-but-not-quite-birds of the troodontids specifically have been marked out as having a particularly high EQ compared to other dinosaurs and thus are almost always credited as the most intelligent of their kin.*

---

* A recent suggestion that *Tyrannosaurus* could be as smart as a chimpanzee based on its EQ was rapidly corrected by the researcher who suggested it and, in short, no, they weren't.

Being one of the less smart ones does not make an animal stupid, though, as surprisingly little brain is required for some complex behaviours and animals such as sheep (an animal rarely credited with many smarts) have excellent memories, and can recall the faces of other sheep a year after they have seen an individual only briefly. So while sauropods might be much less intelligent than other dinosaurs, they should not be written off as dumb and lizard-brained.

Indeed, in recent years the reptiles as a whole, and crocodilians in particular, have undergone something of a reappraisal and are now credited with many more 'advanced' behaviours than previously realised. We see some complex social interactions in green iguanas, for example, and play behaviour (previously thought to be exclusive to mammals), so it is important to stress that reptiles themselves are not stupid.

One recent suggestion that alligators were tool users, collecting sticks as bait to lure in birds that were trying to build nests, has since been dismissed as probably incorrect. This had instantly led to the speculation that if alligators were mentally capable of these actions, then dinosaurs may have been too, and while the idea didn't survive long, it shows that researchers are at least taking seriously the idea that reptiles were capable of such behaviours.

The long and short of all this is that we have a pretty poor idea overall of quite what behaviours dinosaurs were capable of. We don't even have a great idea of what modern reptiles can and can't do, so it's not going to be easy. It is compounded by the problem of getting good measures of brain size or for that matter body size of dinosaurs to even get patterns like EQs available, let alone interpreting them appropriately. A huge amount of possible behaviours are very much on the table for dinosaurs, and we can be certain they were not simply stupid brutes (though some must have been, or at least did stupid things at times).

Social or solitary?

An area of behaviour that has become rather overblown and over-stated is that of social behaviours. A great many dinosaurs have been said to be social, and even have sophisticated interactions between group members, based on extremely flimsy and even effectively non-existent evidence. Doubtless many dinosaurs were engaging in various social behaviours, but the data is very sparse. Before diving into this in a bit more detail, it's worth dwelling on the term 'social' itself, as this has been used at times to cover a multitude of different behaviours.

While it's clear that animals such as chimpanzees and meerkats live almost exclusively in groups with strong interactions between different members of a group, others like cheetahs or wolves are perfectly capable of living alone or together (and will switch between the two readily), and still others such as leopards or moose are generally solitary.

However, many species will appear in larger groups, but without much or any real interactions between them that would make them 'social'. As noted in Chapter 6, wildebeest are famous for their mass migrations in Kenya and Tanzania with up to a million animals moving as some kind of super-aggregation, but at a social level this is made up of many smaller units consisting of a male and a handful of females, and outside of the migration and breeding season, these animals often live alone. In South Africa, though, males don't migrate and in the winter will sit alone holding their territories while only the females and juveniles move. These sorts of behaviours are therefore extremely plastic and variable according to the species and the local environments, and closely related species (e.g., red deer and moose) can do very different things.

Turning to dinosaurs, the evidence presented for sociality is generally from bonebeds with large numbers of animals of a species found together and trackways showing multiple individuals moving in the same direction, and also sometimes large groups of nests. Obviously these only ever capture limited timeframes, and while a breeding season could last several months, for example, there are plenty of solitary species that aggregate to breed for safety and are neither social, nor even spend the majority of their time with other members of the species.

Similarly, even vast groups of dinosaurs from tracks or mass mortality sites need not represent any kind of social grouping, but could be part of a regular migration between feeding sites or forced ones because of poor conditions and changing environments, or may represent only a very temporary coming together, perhaps to breed or exploit a local resource. There are photos online of dozens of grizzly bears together when feeding on salmon, or on a dead whale on a beach, but we wouldn't call them social, however many tracks they might leave in one place.

Massed trackway made by a small group of small
theropods from eastern China. All the tracks are about
the same size and facing the same way, suggesting these
were moving together. Photo by the author.

This is absolutely not to rule out social behaviours; there are too many tracks, bonebeds and nest clusters (and simply too many species of dinosaurs of such colossal variety) to exclude this. But claiming there were complex social interactions of certain species or even entire clades based on this sort of data is often stretching things too far. Tracks of only two of three animals, which could have been laid down hours, or even days apart, have been used to argue social behaviour across an entire radiation of theropods and this is plainly not reasonable. Animals forming groups and even living together was likely common, but it is very hard to demonstrate.

Perhaps the best evidence for this comes from the little ceratopsian *Protoceratops*, which heralds from the Late Cretaceous of Mongolia. In addition to a known cluster of adult animals and one of very young juveniles, I and my colleagues described a pair of subadults and a group of four young *Protoceratops* together. The environment these came from is not considered especially seasonal, so there's no special reason to think these clusters represent only one time of year. If they were coming together to migrate, we might expect all these different ages to be together, and if they only came together to mate, then the juveniles wouldn't do this.

So here we can make a decent case that throughout their lives, *Protoceratops* were getting together in groups. Even this can't rule out that it may have been something that was linked to bad weather, perhaps, or that most individuals usually lived alone, but it's probably the best example we have.

Improving our understanding of social behaviour will come, however. New discoveries of groups of dinosaurs together and greater understanding of these sites will help enormously. Where we do have multiple aggregations of single species, research into the preservation of each site and things like the local pollen will help show if these are all from similar seasons, or if the fossils all represent times of severe drought. If so, this might favour an explanation of migration, but if not it would support these being persistent groups.

It would also be interesting to know if these groups are all male or all female, or mixed. At least some are known to include adults and juveniles, which is itself interesting, though again the composition is unknown. Both elephants and giraffes often have males generally living alone, with females in groups with their young, and it is possible dinosaurs did something similar.

On that note, one other area that will help us to understand such possible behaviours comes from a better understanding of the ecological pressures that affect living animals and how this might correlate with some behaviours. We do know that as levels of predation increase, species that can live in groups or alone tend to come together more in response to the increased threats. There is also an interesting correlation between body size, mouth size and sociality in antelope, which has potential for explaining what dinosaurs may have done.

In short, large animals need a lot of food and can't rely on highly nutritious things such as fruits and shoots, and they have long digestion times allowing them to take rougher foods. This also makes them bulk feeders with wide mouths. If they live in open environments like grasslands, then it is difficult for them to conceal themselves from predators, so forming loose aggregations or even large groups would be beneficial to help them spot things coming. If on the other hand they live in dense environments, then living in a small group would give them some benefits of extra eyes and ears, while still being able to hide.

Small animals however, can't digest rough food, but are small enough to exploit nutritious, but rare, foodstuffs like buds. They have small mouths to select the most nutritious bits of plants to eat, but they live alone or in pairs. This allows them to remain concealed and also to defend a small territory and keep competitors out. Thus, based on size and mouth shape, with some information on where they lived, this could help to reveal if dinosaurs lived alone or in pairs, in small groups, or potentially in large ones.*

Getting the message across

If animals did live in groups, then one important aspect of that is how they communicate and even establish dominance. As with the previous chapter, many features that serve as sexual displays and signals can equally function as dominance indicators in a social setting. Just as a big set of horns and a large colourful frill on *Triceratops* would be a show of size and quality of a male appealing to a female, it would also show any other individuals that this was an animal not to be trifled

* This has been applied to dinosaurs, though with limited success because of our lack of detailed understanding of their feeding ecology, but it's still a great example of how an anatomical feature such as mouth shape could potentially be linked to social behaviour of fossil animals with the right data.

with, and so any competition over access to water or the best feeding spot would go to them. Thus all the vast array of horns, rfills, crest, plates, feathers, calls and other features could have been used equally as social signals as sexual ones.

Indeed, very often we see in modern animals that sexual signals function in social dominance. These can similarly be used when things escalate, too, and horns and fangs in particular could be turned on another individual purely for access for some limited resource, rather than necessarily for trying to access a mate.

These features would also be used when faced with members of other species that could be a threat. Naturally predators will be a constant issue for any herbivore (and many young or small predators), but other species will be competitors for various resources, be it food, water or access to something like a burrow or good nest site. We tend to overlook these sorts of interactions, as they may be rare or there is not much apparent conflict going on, since if a 30-ton sauropod wanders over to drink, the 300-kg ornithischians are going to give way without much of a fight. But even herbivores can get into major engagements, and there are videos online of elephants in Africa fighting with rhinos and water buffalo.

This would likely have been a very minor part of the behaviour of many dinosaurs, but stegosaurs might well have used their tail spikes in a spat with a small sauropod, and hadrosaurs could have been charged by ceratopsians or pachycephalosaurs. I'm not aware of any fossil that would give evidence for these kinds of interactions, but that would have been rare, and it may only be a matter of time before we find a specimen that would show something like this.

It is in the realm of predators and prey that the most obvious cases for defence (or indeed offence) would come from potential prey species. There is an *Allosaurus* known with an odd hole in one of its tail vertebrae, which has been suggested to have been made by the tail spike from a *Stegosaurus*, but, while plausible, this has not been widely accepted. Similarly, there are a number of Cretaceous tyrannosaurs that have healed injuries on their lower legs and feet.

While no formal analyses have ever been made of this, I can't help but find it more than a coincidence that broken shins seem to turn up regularly in the one group of large carnivores that lived alongside the tail-club-wielding ankylosaurs. There are other possible explanations for this, but it's an intriguing possibility, and it would be odd indeed if ankylosaurs didn't try to defend themselves with their most obvious weapons against their most likely antagonists.

Herbivores may well have defended themselves with some of the methods suggested at various times. Ankylosaurs would be well advised to squat down where their unarmoured undersides would be even better protected from attack and making them even harder to get at, while large groups (assuming of course there were large groups) may well have kept young animals to the centre or the rear, and the largest animals may have charged out to face threats. However, these ideas, while reasonable and adopted by various living groups, are almost impossible to determine, though appropriate footprints may one day reveal such stories.

Dining out

When it comes to carnivorous behaviours, we do have excellent evidence for some of what was going on. Various theropods have been found with bones of other dinosaurs inside them, which tells us what they ate, and we even have various herbivores with theropod teeth wedged into their bones and signs of healing around them showing they survived an attack. We also have huge numbers of bones showing various marks on them from the teeth of theropods as they fed, though it's dangerous to infer from these any kind of hunting patterns.

After all, pretty much all carnivores will scavenge when food is available and except under unusual circumstances, where we can reconstruct the history of the carcass before burial, it's basically impossible to tell if those bites were made by an animal that had killed its prey, or was only scavenging. At least in a few cases, the latter is pretty obvious, such as when we see marks left by very small theropods on a big dinosaur, but overall we can't say much more than that theropods were both active hunters and opportunist scavengers – and this is annoyingly trite as a revelation about their behaviour.

We can, though, begin to delve a bit deeper when considering the anatomical specialisations of various groups and the patterns of bites and stomach contents, and even the fossil record as a whole. There is a strong bias against young animals being found, as their weaker bones do not fossilise as well as big ones, and they are small and hard to find. But on the other hand, juvenile dinosaurs should have been exceedingly numerous based on dinosaur eggs, and yet they remain extremely rare.

Notably, however, most incidences of theropod-injured dinosaurs and the stomach contents of theropods are those of young dinosaurs. Among living species, juveniles are almost universally preferred as prey by carnivores since they are generally numerous and less experienced at avoiding predators, and usually have weaker defences (e.g., lacking horns).

All of these lines of evidence tie together to suggest that the reason young dinosaurs are generally rare in the fossil record is that they were being eaten and so were less likely to be buried and preserved as fossils. One last fact also plays into this: when we do find good specimens of juveniles they are very often in groups of multiple individuals, even when the adults of that species are never found together. As noted above, grouping is a common defence against predation, so a natural explanation for juveniles aggregating even when adults do not is predation risk.

Even so, not every theropod would be targeting juveniles, and in addition to old, injured or sick adults, even healthy ones would be attacked at least on occasion. Quite how the various theropods did this is very unclear, though we have some indications. Pretty much every modern large and active predator takes one of two approaches, sneaking in as close as possible and making a quick dash to attack, or pursuing prey over long distances to wear them out.

In the case of tyrannosaurs, for example, we see adaptations for efficient travel in the proportions of the legs, and modifications to stiffen the joints between the bones of the foot, and coupled with the fact that it's probably hard to hide an animal that can stand several metres tall and weigh multiple tons, then a sneaky approach is not likely to be an option very often. On the other hand, plenty of the smaller theropods would appear to be quite capable of hiding well given their size, and are lightly built with adaptations for being quick.

A huge number of species sit between these extremes and it can be very hard to know quite what they might have been doing. The quintessential Jurassic theropod *Allosaurus*, for example, does not appear to be especially fast or a particularly good long-distance runner, and while it's clearly too big to sneak around behind small bushes or ferns, it's a lot smaller than many theropods and could at least potentially have taken cover in trees, or used the cover of night. That said, if it was predominantly hunting juvenile sauropods, few of these would have had any real turn of speed on them, so perhaps a lack of oomph was no real handicap.

Without knowing what they were after and what the local conditions were like in terms of cover, light, heat, how good the senses of their typical prey were, and whether or not those were living in groups or in dense cover, it becomes difficult to make any real assertion over how many theropods may have hunted. However, many of these issues are things we are getting better at working out and new fossils will provide ever better evidence of active predation attempts, so while right now we have little idea on this area, it is likely to improve.

One aside that is worth also considering is the idea that theropods hunted in groups (most often called 'packs'). Once again this is plausible, but evidence for this is extremely thin on the ground and really comes down to a couple of mass mortality sites for some theropods. As noted above, even this sort of data is not especially convincing that these animals normally lived in groups; even those modern predators that are highly social (including lions, wolves and hyenas) often or even mostly hunt alone. Even when hunting together, it is often a case of all rushing in at once and hoping someone gets lucky and grabs something, allowing the others to pile in, rather than some organised and structured plan.

While lions can lay ambushes, the idea that various theropods (most notably dromaeosaurs and tyrannosaurs) had complex hunting strategies with specialist roles for individuals is completely unsupported. Even some amazing complex trackway would be hard to interpret as reliably showing this, so it remains firmly in the realm of 'maybe', although we can't say so with any real confidence.

Once a predator has caught up with a potential meal, it needs to bring it down. Clearly most theropods would have used their teeth, and in the case of the tiny-armed tyrannosaurs and abelisaurs that was pretty much their only option. For many of the theropods closest to birds, these also had very long arms with large claws, and there are the famous 'killer' claws on the feet of dromaeosaurs and troodontids, too. These were probably used less for slashing or even clambering onto and holding on to big dinosaurs, as once believed, and more for puncturing and gripping small prey.

Again, though, we are left with an odd consortium of animals that fit neither of these profiles, having medium-length arms with moderate-sized claws whose function is far from clear, as covered in Chapter 9. This is an area ripe for further research and a better understanding of how these arms would work (how strong they were, the exact range of motion, the 3D shape of the claws, and so on) will feed into this. A better understanding of what they were trying to predate upon would also help us to work this out, but it remains an area of great uncertainty.

Rise and shine

One other major aspect of the interactions between predators and prey and even the basic biology of dinosaurs is their daily rhythms. We tend to picture them as being animals of bright sunlight and tropical climes, but there's no reason to think many of them were not only active in low light, but actively nocturnal and shunning the sun. If nothing else, large theropods would find it considerably easier to conceal themselves at night and any dinosaur over a few tons in weight that lived in a hot climate would be at a massively reduced risk of overheating out of the sun.

In addition to metabolic arguments and local climate, we can also look at the senses of dinosaurs, since these would have evolved in conjunction with their normal activities. Several dinosaurs have been suggested to be nocturnal, based on the size of their eyes, but this is difficult to assess in detail, since while large eyes do correlate with nocturnal or crepuscular habits, they are also linked to predation. So are the large eyes in tyrannosaurs because they are predators, or nocturnal predators? We can, though, add in data from other aspects of their senses.

Returning to the large-brained troodontids, this is one example where we do have supporting data to say something about their activity cycle.

For some well-preserved dinosaur skulls, we can scan them with powerful X-rays and reveal something of the internal structure of the ear. This gives an idea of what kinds of sounds they were attuned to (high or low pitched), but in the case of some troodontids we also discover something more unusual: the ears are asymmetrically positioned in the skull with one higher than the other. This is an anatomical oddity that turns up in a much more familiar lineage of small, feathered, predatory dinosaurs: owls. Such asymmetrical ears help animals estimate the distance to a noise very accurately and so it makes a lot of sense when hunting at night. Troodontids also have large eyes and, coupled with the arrangement of their ears, this makes them good candidates for being very active at night.

Similar scans of skulls can also reveal something of the structure of the brain inside. As these are structured in similar ways, it is possible to look at the various different lobes and see (roughly) how much of the brain is devoted to processing visual or olfactory cues.* Again, these are early days, but we have the methodology in place and there are plenty more skulls out there that can be scanned to build up a picture of how dinosaurs sensed their world, which can only be an important stepping stone to understanding their behaviour.

There has been perhaps more research into dinosaur behaviour in the last decade or so than over the entire previous century and a half combined. A better appreciation for the behaviour of modern animals (especially reptiles) and the evolutionary and ecological links to certain activities means we have a much better basis for restoring what dinosaurs were doing. Add to that new data from the skulls to determine their senses and intelligence, and ever more fossils that give insight into where they were living, what they were eating and how they communicated, and this is a field ripe for a marked improvement.

* The legendary incredible sense of smell of *Tyrannosaurus* comes from the idea that it had an enormous olfactory processing part of the brain, but it turns out that this mistook some of the space of the sinuses as being part of the brain and thus greatly overestimated the acuity of their sense of smell. It certainly wasn't bad, but it was not the super-smeller it has been claimed.

While there is enormous uncertainty over a lot of the basics of dinosaur behaviour, and worse, much has been overstated that can't be well supported by the evidence, this is an area that will be developing rapidly in the foreseeable future.

14

Ecology

IT IS OFTEN difficult to work out the ecology of many living species in detail. Animals roam far and wide in their daily lives and observing enough of them for long enough to determine what they eat, how much of it, what triggers them to migrate, which species they avoid or are their biggest threats, who they compete with for food or space, and a whole host of other facts, is extremely hard to do. Even watching numerous individuals 24 hours a day for a year, it would still not be sufficient to work out all the nuances of their behaviour in response to changes in humidity or temperature, or the scent of a predator, or why they avoid some plants on one day and devour them a week later.

As may be imagined, therefore, if we are still often struggling to say whether or not the dinosaurs migrated, whether they were nocturnal or diurnal, when and where various species lived and so on, then reconstructing their ecology beyond some of the most basic details will be challenging to say the least.

Since this data is so hard to come by, it may not be a surprise that it has been an area of research that has rather lagged in dinosaur studies. Beyond basic data such as that sharp teeth are found in carnivores and big groups of dinosaurs found together may have lived in herds, little thought or justification was given to most interpretations of dinosaurs as animals that lived in real environments with other animals and plants, and the abiotic factors like the climate. So a combination of little data and little thought leaves our understanding of their ecology lagging behind that of many other areas, though with the exciting flipside that there is still so much to discover here.

Dinosaur diets

The most fundamental part of an animal's ecology is what it eats. Here at least we have a very good idea for most dinosaurs, since the basic tools for feeding and processing food, their teeth, usually point very clearly to carnivory or herbivory. More than this, we can often work out some important details and subtleties of a dinosaur's diet from data such as stomach contents or coprolites, which can directly show what they were eating. Further detail comes from tiny scratches and pits on the enamel of their teeth (termed 'microwear'), which in some cases can be correlated to certain types of food that leave distinctive marks, such as pine needles.

More obvious damage to teeth can also reveal interesting aspects of how animals fed; for example, the famous sauropod *Diplodocus* shows massive wear on the back of their teeth inside the mouth, but little on the top where it would be expected. This is interpreted as them feeding by taking small branches and fronds into their mouth and then dragging the head back so that the leaves would be stripped off. That action would mean that the majority of the work, and therefore the wear, took place on the back of the teeth, explaining the wear there that is limited elsewhere on the enamel.

The mechanics of the movement of the neck align with this interpretation, so here we can be confident we have a good idea of one of the main feeding activities of this animal. Similarly, some basic details on other species and groups can be told from their anatomy. The low-slung bodies of the ankylosaurs must have been feeding low to the ground, and they often have surprisingly small mouths, pointing to selective feeding of nutritious shoots and buds. In contrast, the sauropod *Nigersaurus*, while also being a low feeder (based on its short neck), has an extremely wide mouth that looks something like a vacuum cleaner attachment, and must have been taking in as much as it could with such a gape suggesting it was an indiscriminate feeder.

However, there remain various groups whose diet remains uncertain through to downright controversial. The troodontids, despite their close kinship to the carnivorous dromaeosaurs and earliest birds, have several times been suggested to be herbivorous. They do have clawed hands and feet, which would appear to make them predatory, like their nearest relatives, but these could potentially have been used in spats between individuals, or for these small animals to fend off threats from larger theropods; so the claws don't rule out herbivory (while obviously functioning rather differently, check out the huge claws on sloths and pandas, for example).

The support for this idea comes from their teeth, which have unusual giant serrations on them, quite unlike those of other theropods and reminiscent of various herbivores, including the iguanodontids. Coupled with the information we have on the brains, eyes and ears of various troodontids, this gives an odd melange of traits that could point to a smart herbivore with amazing senses, or a small carnivore with odd teeth, or perhaps an omnivore capable of catching small things or processing plant material when the need arises.

Similarly, the diet of some of the oviraptorosaurs is also unknown and debated. Their name derives from the first discovered genus in the group *Oviraptor* ('egg thief'), which was interpreted as eating the eggs of other dinosaurs, when we now know the reason the skeletons were often associated with eggs is that they were brooding. Various early forms have a fairly blunt beak, which looks ill-suited for a carnivorous diet, but more importantly, many specimens are found with numerous small stones in the stomach called gastroliths, which would have served to break up seeds or similar tough plant material.

These would function like the stones do in the crops of many modern birds, including chickens. However, among some of the larger and later species (including *Oviraptor* itself), there's no evidence of gastroliths, even in very complete and well-preserved specimens, and their beaks are sometimes rather sharper. That leads to the suggestion that some or many of these were carnivorous, but there's no good data either way to support this. A specimen with stomach contents of plants, or bones, or even both, would help give some weight to these various hypotheses.

Even when we have a good idea of what things were eating and also how, there are the major curiosities and unknowns. In particular, the huge range of sizes (and even changes in shape) that dinosaurs underwent as they grew meant that, ecologically speaking, they often occupied multiple different niches as they were growing. Large tyrannosaurs, for example, were very leggy animals when young and probably fast on their feet, but adults were much heavier built and had larger heads with stronger bites, and were better suited to long-distance pursuit than a sprint.

That clearly suggests a change in the style of hunting and also the kinds of prey they were after, but quite how they transitioned from one to the other isn't known. Similarly, young sauropods simply cannot have been doing the same thing ecologically as large adults, as they would not have had the reach to get up into trees, or the digestion time to tackle most tough plants effectively, so they (and many others) must have undergone ecological changes as they grew.

These changes have massive implications for what dinosaur ecosystems were like. After all, if the populations of most species were made up of juveniles (and remember, a young sauropod could still be an animal of a couple of tons), then animals would be occupying and using the landscape very differently to one composed mostly of adults. This would also then change what it meant in terms of population density (you can have a lot more small ones than big ones), and also how predators acted (if most of your potential prey was much bigger than you, then social hunting suddenly looks very attractive). Then, of course, there are the layers of interactions and competition between species.

Palaeontologists have looked at these kinds of issues in terms of the relative feeding heights and diets of various dinosaurs in certain formations, to see how they may have separated themselves out. In ecology this is termed niche partitioning and relates to how competition between species will tend to drive them into exploiting different ecological resources (such as food and water, but also nesting areas) and there's a fairly clear pattern, with some ecosystems having very low feeders, other animals taking food at intermediate heights, and others capable of getting food from high places like trees.

A high browser, such as an adult *Brachiosaurus*, is therefore probably not in much competition for food with *Diplodocus*, but a half-sized one might be reaching to a very similar height as *Diplodocus*. Of course, they may well be specialising on different foods, or living predominantly in different areas, but there will naturally be some overlap conflict (and these are far from being the only sauropod species in the area), and what that means for each other and the environment as a whole is almost impossible to predict.

If we had a great deal more information about a typical dinosaur ecosystem, we might be able to say rather more about their biology. Our understanding of ecological theory is good and growing, but the problem with ecology is that it is so incredibly complex, with all the interactions of different species (plants and animals, herbivores and carnivores, parasites and diseases), weather, seasons, climate, and other things like soil quality, humidity and underlying geology.

Sum together all those abiotic factors and thousands of living species and all their interactions, coupled with the fact that things will be in constant flux even on a daily or annual basis, let alone across half a million years, and it should be clear why we have such uncertainties about how even well-studied living ecosystems function. Try to do that with a fraction of the available data, and with huge gaps in the data we do have, and it's inevitable that our grasp of dinosaur ecology is somewhat weak.

Ecosystems

Even well-studied ecosystems can be a cause for confusion. The famous Morrison Formation covers much of the Midwest of the USA and is home to numerous famous dinosaurs (*Diplodocus*, *Brontosaurus*, *Brachiosaurus*, *Stegosaurus* and *Allosaurus* among others). It has been subject to a vast amount of study, with so many specimens of so many species, but it is proving resilient to consensus. True, it is complex, with numerous different fossil localities, and it covered a huge area and a long period of time, so we would hardly expect it to be homogeneous, but there's major disagreement about whether or not it was a mainly wet and forested ecosystem, or a dry and open one.

That alone can shape our understanding of whether probably supported large numbers of animals or few (of which more below in a minute), and therefore whether or not species might be forced into regular migration to obtain food. More and more evidence in this case is pointing to the wetter end of things, but it's taken a long time to get here given the amount of work done in this area.

It is not helped by the inevitable fossil biases – the Morrison was made famous by the diversity of sauropods and other dinosaurs, and also the huge mass mortality sites preserving thousands of bones. Major droughts apparently led to the burial of vast numbers of bodies together, but then this presents a problem. This is a major source of data for palaeontologists, but if that's being produced under abnormal conditions that have drawn or forced animals together from a wide area, it may give a very misleading picture of what lived locally and in what kind of numbers.

The Morrison is not the only dinosaur-dominated rock formation that is difficult to interpret. The various beds of the Yixian Formation in China, which has produced so many wonderful feathered dinosaurs, are absolutely dominated by small species. Not just small dinosaurs (which are very plentiful, in contrast to almost all other formations), but also amphibians, mammals, birds, lizards, and others are common too. Large dinosaurs are present but notably rare, with only a few sauropod bits and some iguanodontians found, despite decades of intensive excavations. The nature of the formation being composed of super-fine volcanic ashes is ideal for preserving small animals and soft tissues (hence the plethora of feathered dinosaurs), but then begs the obvious question: is this normal?

We know there are biases against small animals being preserved in most places, so, on the one hand, the Yixian may give us a better idea of the kind of numbers and diversity of small dinosaurs we might normally expect but simply can't find. However, the rarity of large ones under exceptional preservation conditions suggests that this is a far from normal ecosystem, or that large animals are for some reason not preserving here. Either way, it's impossible to know and so limits our ability to try and use this data to make a better informed impression of what a normal dinosaur ecosystem was like.

Some are difficult to understand because they are so obviously abnormal. Perhaps the most notable case is the Kem Kem of North Africa, home to the famous giant theropod *Spinosaurus*, which was probably eating a lot of fish there and spending time in water. This controversial animal tends to suck up the headlines and so what is often overlooked is that the formation as a whole is very unusual.

In addition to *Spinosaurus*, there are also several other large theropods, such as *Carcharodontosaurus* (these being very much fully
terrestrial), but also various giant crocodiles. So on land or in water
there were lots of large carnivores, but there is little evidence of
other dinosaurs. There are some around but they are not common,
and an ecosystem with so many large predators must have had
something for them to eat.

The answer 'fish', which are themselves very plentiful, seems a
good one, but becomes rather unsatisfying given the numbers of
crocodiles and the lack of fishing adaptations in all but *Spinosaurus*.
Clearly this was a functioning system and the large theropods were
eating something (probably large herbivorous dinosaurs), but why
these animals are not preserved is far from clear. Perhaps the
herbivores spent most of their time elsewhere, but then that hardly
explains why the theropods would hang around where there's no
prey.

Another odd one is the Solnhofen limestones of southern Germany.
Most famous for producing the 'first bird' *Archaeopteryx*, these fossil
beds mostly preserve the various local marine faunas, including fish,
turtles, shrimp, horseshoe crabs and numerous pterosaurs. The area
was made up of a series of lagoons and an island archipelago, so it's no
surprise that most species found there either flew or lived in water. In
addition to *Archaeopteryx*, there are some other small feathered theropods known, several of which could fly. Presumably this is why these
animals are relatively common in these rocks (though still much rarer
than the pterosaurs), since while they were foraging on land or perhaps
occasionally shorelines, they would from time to time have flitted
between islands or been blown out to sea, and ended up becoming
buried and fossilised.

Of more interest are the compsognathids that have been found
since these could not fly and presumably rarely, if ever, entered the sea
deliberately. But while there are these small theropods, there are no
other dinosaurs at all. Are the islands so small that they could only
support a small population of small theropods, or are the herbivorous
dinosaurs simply not turning up?

Maps of the Solnhofen reconstructed from fossil reefs and other
data suggest that the mainland was not too far away and certainly some

islands were fairly extensive, so one would think that at least occasionally some other dinosaurs would make it out this far, either living locally or getting swept out to sea in the seasonal storms. For now, though, it remains frustratingly uncertain, and without more discoveries it may be hard to work out quite what was going on here.

Other things we can be more certain of are that there would have been plenty of strange and unpredictable relationships between species. Even if they are a very rare component of an animal's normal life, odd things do happen. A couple of years back, some hadrosaur coprolites were described that contained remarkably high levels of wood and, much more surprisingly, fragments of crustaceans. It is hard to imagine that hadrosaurs were snacking on crabs or shrimp, but these bits of shell turned up in multiple specimens and were from decent-sized arthropods, which means they would have struggled to eat these by accident.

Plenty of living animals we think of as dedicated herbivores will take a bite of meat if it's available (there are some fascinating but rather sinister videos online of domestic cows hoovering up baby chicks), so this is perhaps less odd than it might at first appear. It's obviously a challenge to our assumptions and how we picture these animals, but such incidences are rarely recorded, so it is hard to know if they were common or rare.

Specialisations and interactions

We can easily over-hypothesise, reconstruct or over-interpret small bits of data into huge assertions, and while that may be fun to do, it's not very informative. What is more appropriate is to look at some of the more common patterns of ecological interactions and relationships that would be almost inevitable in dinosaurs and their possible consequences. While we have already looked at the basics of carnivory and herbivory (and at least touched on the possibility of omnivory), this glides over plenty of subtlety.

One of the most fundamental questions is whether a given species is a specialist or generalist. This is very much a continuum rather than a binary case, but the two extremes are notably different. Specialists have a very limited niche, typically related to their food source, but also potentially including their range or tolerance of various conditions. A good example would be one of the various moths that have evolved intimate relationships with plants, such that they get the vast majority of their food from a single species. They are highly specialised and cannot survive without that resource, though they often don't have much competition from other species.

Generalists on the other hand are a jack-of-all-trades. Take leopards that live on savannah, mountains, deserts and in rainforests from South Africa to Indonesia and will take anything from sparrows to baboons and antelope as prey. They can switch easily from one habitat or food source to another, but they will face competition from the various local specialists of any given area.

Some dinosaurs would be more at one end of this continuum and some the other, and we can predict that generally the larger ones would be generalists. For both carnivores and herbivores, the ranges they would need to cover to find food and maintain a breeding population would be so large that it would cover numerous different terrain types and local variations in climate and available food, so they would simply not be able to be specialists.

However, at the smaller scale for young animals or small species, being a specialist is much more viable. That is very hard to determine, though, and while doubtless there were specialists out there, we have yet to find anything with such specialisations to their anatomy that we can make a reasonable inference of a highly specific diet or activity. Perhaps the closest are the smallest alvarezsaurs, which are insect-eaters and may even have specialised in things such as beetle larvae and ants, which is a narrow diet, but hardly specific to eating only one or two species.

Another major ecological phenomenon that we can expect some dinosaurs to have produced is the 'top-down' effect. To put this into context, we must first look at the reverse – the unsurprisingly named 'bottom-up' effect. This is where populations and entire ecosystem structures are controlled by the available nutrients and productivity of an ecosystem. Places such as deserts or mountains have major limitations on what plants can grow, with the effect that this will then place a critical limit on the population of herbivores, and therefore also carnivores, and so on. Top-down,* however, comes from the idea that it is predatory animals that are ultimately controlling the makeup of a given ecosystem, with their actions manipulating the number of herbivores and/or the makeup of their populations and corresponding behaviour.

In the case of many large theropods, they would seem well-positioned to do this, though in the sauropod-dominated faunas this could get very complicated. It's not hard to imagine that *Allosaurus* had a major effect on the population structure of the sauropods by taking out numerous juveniles, but may have had little effect on the adults, whose dietary needs could easily shape whole landscapes. Thus some complex trade-offs likely existed, given the abilities of various carnivores to tackle certain sizes or types of herbivore. It would be difficult to demonstrate, but as we learn more about the behaviour and prey choices of theropods, it may be possible to understand better which ecosystems contained theropods that were probably acting as top-down controllers.

Ecosystems are dynamic places and while over long periods of time they may be basically stable, such that a rainforest may be present on the same spot with many of the same species for tens of thousands of years at a time, the details will constantly shift. Various other regulators of populations would have been acting, such as changing climates, events like droughts and floods, or invasions of new species, which could have had some profound effects.

I have focused here on the interactions between dinosaurs with each other and with herbivores eating plants, but it would go far beyond this in reality. Small juveniles would be vulnerable to being predated on by the larger lizards, mammals, and pterosaurs; fungi would break down dinosaur faeces; many herbivores would have had important symbiotic bacteria in their gut flora; and doubtless some plants would have been

---

* This comes from the idea that predators are at the 'top' of a food chain. It is worth noting, though, that the endlessly repeated idea that animals such as *Tyrannosaurus* (or indeed modern carnivores, like lions) are 'top' or 'apex preda-tors' is one that is misplaced. This is often used as a synonymy for whatever is the largest predator in an ecosystem, but in ecology the term is generally reserved for those predators that are not only at the top of the local food chain (so nothing normally eats them), but *also* predominantly eat other predators. So, great white sharks that are eating sealions and large predatory fish are top predators, but lions that eat gazelle and tyrannosaurs eating hadrosaurs are not.

toxic, while others may have relied on dinosaurs to spread their seeds or even for some to act as pollinators. All of these and more would act together to make up complete and functioning terrestrial ecosystems, and dinosaurs would have been but one (if major) part of them.

Parasites and diseases

We all know *Jurassic Park* was predicated on the idea that ancient mosquitoes sucked up dinosaur blood before being preserved, and while the idea that DNA could be extracted from these is a fiction, there absolutely were mosquitoes, ticks, leeches and other bloodsuckers around in the Mesozoic. It would have been quite extraordinary indeed if they did not habitually feed on the biggest and most numerous animals in their respective environments. Evidence for diseases, parasites and infections in dinosaurs is mixed, but like all living vertebrates they must have had plenty. Numerous dinosaurs show a wide variety of unusual bone textures and changes that are the hallmarks of an infection serious enough that it afflicted the bone.

Various other oddities such as holes in the jaws of various tyrannosaurs have also been noted as looking like those produced by a nasty protozoan infection called *Trichomonas*, which infects modern hawks and raptors, and may have got them too. There must have been all kinds of things like flukes or worms of various types in the blood or major organs, as well as bacterial and viral infections. Getting any good evidence for these is nearly impossible, however, let alone an understanding of their spread and degree of infection and virulence. Certainly some insects and other blood feeders would have been vectors, but that's about all we can say for now.

Doubtless dinosaurs would have experienced occasional outbreaks of serious infections, or fallen victim to mass poisoning from blooms of toxic algae in water holes or similar incidences. Most of the mass mortality sites we have can be attributed to floods or droughts, but it's possible that groups got overwhelmed by some infection or other at times. It was once suggested that, in addition to the obvious ecological disaster of the K–Pg event, dinosaurs were already under serious population pressure from diseases.

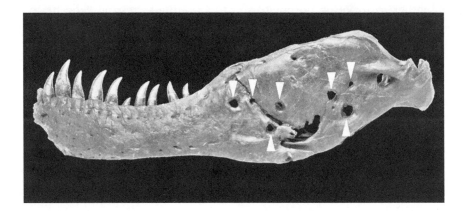

Holes in the jaw (indicated by white arrows) of a
T. rex suggested to be caused by an infectious protozoan.
Image from Wolff, E.D.S., Salisbury, S.W., Horner, J.R. and
Varricchio, D.J. (2009). 'Common Avian Infection Plagued
the Tyrant Dinosaurs'. *PLoS ONE*, 4(9), pp. e7288.

This is near enough impossible, though; remember that an infection would have to affect billions of individuals across all seven continents (and how would it spread across oceans?), with no natural immunity existing or evolving. Even the very worst pandemics that we have seen are rare in affecting only a few species and typically those are ones already vulnerable for other reasons, since the mortality rate of even horrific illnesses is often well below 50 per cent. By the end of the Cretaceous, the sauropods, theropods and ornithischians had been evolving separately for over 150 million years, so it seems unlikely they all had such a similar physiology and immune systems that a single infection would be equally devastating and fatal to all.

Dinosaurs would certainly have had their share of death by disease, but there's no reason to think it would have anything like this drastic an effect. More reasonable would be that some major new disease did help shuffle off a good few species and even the odd lineage. A breakout in some ecosystem could well have removed the most numerous predator or herbivore and drastically changed the local dynamics (at least for a while), leading to some interesting changes, though sadly of the kind we will never be able to reconstruct.

Overall, though, dinosaur ecology is mired deep in the problem of a lack of data. The study of modern ecosystems and the interactions of species and their environment is one based on the huge numbers of intertwined pressures that are exerted in all manner of directions and that are constantly changing. Looking back tens of millions of years with only a vague sense of what species were present and what the general climate may have been like, let alone details such as the makeup of populations and which species they fed on, or how much water they needed, leaves us painfully short of data.

Still, as with so many areas, an increasing appreciation of the kinds of dynamics that must have been present (top-down effects, diseases, generalists versus specialists and so on), and more and more information about the fundamental biology of dinosaurs from their diet to reproductive rate and physiology, means that our understanding improves. This area may take longer than many others covered in this book to yield major strides forwards, but there is inexorable and incremental progress here too and our understanding of dinosaur ecology, while patchy, is improving.

15

Dinosaur Descendants

Around ten thousand dinosaur species are still alive today in the form of birds and it would be remiss not to include something on this remarkable clade. I have generally sidelined them here, as I did want to make this book primarily about 'classic' dinosaurs. It would have been too easy to be constantly distracted by the fact that almost every chapter could have focused on the transition from the thero-pods into birds and been every bit as long, but that would have produced an entirely separate and different work.

Outside of the origins of humans, there is probably no field in palaeontology that has had as much attention in the last twenty-five years as the origins of birds. This area has long been one of lively debate and discussion in the scientific literature. The first discoveries of feathered dinosaurs in China in the late 1990s came off the back of increasingly strong anatomical evidence for a relationship between birds and dinosaurs in the 1970s and 1980s, but it has grown and grown since.

The change from terrestrial dinosaurs to true birds is now one of the best documented major transitions in evolutionary biology, in terms both of the number of fossils and the amount of research on it. Even so, there remain some major questions about this transition and gaps in our knowledge, and as a result there are things we don't know and that are yet to be resolved (and some likely never will be).

The ups and downs

Perhaps the biggest of these is the very essence of getting airborne and the question of whether powered flight evolved from animals that lived on the ground taking wing, or those already gliding from tree to tree eventually converting this type of limited flight into flapping. These two positions are generally known as the 'ground up' and 'trees down' hypotheses, and both have had their supporters and much research directed at them.

There are other versions of this fundamental separation, but these are certainly the two most researched, credible candidates. The lack of resolution on this has come from a number of issues, not least an absence of key specimens that might reveal a transitional form capable of limited flapping, or an obvious bird precursor that was definitively good or bad at climbing and might help resolve the conundrum.

One particular issue is that there has been little attempt to reconcile the two arguments, since even an early flapping terrestrial animal would surely have been using its limited flight to get up into and move around in trees, and an early glider that could barely flap would sooner or later have needed to get back up to a good height from which to launch itself next time. In short, it is hard for me not to see these as two sides of the same coin, and the fact that there's little resolution on which of these is best supported by the evidence may be because both are relevant.

The other aspect of this is that there's no guarantee that powered flight evolved only once in the dinosaurs. While it's more than likely that the direct lineage to birds had only a single origin, at least three different groups of dinosaurs (the troodontids, dromaeosaurs and scansoriopterygids) all apparently evolved various gliding forms independently of each other and the ancestors of birds. Since all of these are very close relatives and are known from the same geological formations, we are in a position where we have multiple animals with similar evolutionary histories and anatomical features, and apparently similar evolutionary pressures, all living in the same environment.

If ever there was a condition for several different groups to evolve powered flight at the same time, this might be it. Of course, we don't know if that was the case but, if so, it may also explain some of the contradictory evidence in establishing quite what happened as the birds literally took wing. Several recent studies have provided support for the idea that various non-bird dinosaurs were indeed capable of true powered flight, and it may yet be shown to be the case that there were multiple independent (if very closely related) groups of feathered fliers in the Jurassic and Cretaceous.

Indeed, the very origins of birds at the most detailed level are still something of a mystery. As with the origins of dinosaurs as a whole, we find ourselves in a position where a glut of data actually reduces our ability to answer a question definitively. Groups that are starting to diverge will look very similar to each other, so when you have very large numbers of them – and in the case of so many small, feathered theropods, they are mostly smushed flat into the rock – the details necessary to help separate them out, and understand exactly who is related more closely to who, either haven't actually evolved yet or are simply not visible.

We know the origins of birds lie somewhere around or between the troodontids and dromaeosaurs, but which of these three are most closely related to the other, and when that split occurred, is currently rather uncertain. The exact details of these relationships bounce around a little with each new discovery or analysis.

One aspect of this lack of consensus is that several interesting taxa move around the tree quite a lot and with some dramatic potential implications. The bizarre *Balaur* comes from what is now Romania, but was an island archipelago 70 million years ago, and is famous for its unusual dinosaur faunas with all kinds of island oddballs. Originally *Balaur* was described as a mid-size dromaeosaur (about 2 metres long) with a double sickle-claw on each foot. However, more recent studies suggest that this was an early bird and, if so, while its anatomy still remains remarkable, it would be the largest known bird from the Mesozoic and an early example of the evolution of flightlessness.

Similarly, the legendary *Archaeopteryx* – so often regarded as the literal 'first bird' – is now both rather too recent to be considered for that title (since it's Late Jurassic in age and the birds must have split off some time in the Middle Jurassic) and in multiple analyses has been recovered in different places in the family tree of dinosaurs close to the origin of birds and even closer to the dromaeosaurs and troodontids. If that is the case, then either powered flight evolved multiple times, or possibly even powered flight evolved at the origin of this branch of dinosaur evolution and then multiple groups all lost this trait.

Getting a better resolution on this part of the dinosaur family tree is likely to be difficult, given the gaps and how similar many of these species and groups are to each other; but it is an area of much research and it's likely to firm up towards a single consensus sooner rather than later and help to answer these questions, or resolve on which side these various possibilities reside.

Certainly the lineage that would give rise to modern birds was around by the later part of the Middle Jurassic around 150 or so million years ago. This means that birds lived alongside dinosaurs for the thick end of 100 million years and were not some last-minute innovation that were only around at the very end, or even sprung up as a result of the extinction of the other dinosaurs. The earliest birds (well, to be specific without trying to get too technical, these should be called stem-birds) were still very dinosaur-like. They had long bony tails, large claws on the separate fingers of the hands, teeth in their jaws, and lacked the complex anatomical architecture of the shoulder and pelvis that mark out modern bird skeletons so clearly.

There was not a simple pattern of change here, with various early groups under different evolutionary pressures doing different combinations of things. The recently discovered *Jinguofortis*, for example, manages to combine several features seen in various different lineages of early birds, and shows that there was no simple succession of changes from the 'dinosaur' body plan to the 'bird one' with a progressive shift from one to the other. Teeth were lost several times in different lineages to be replaced by beaks (and at least one species had both), and working out the exact pattern of these shifts, and in particular what evolutionary and ecological pressures may have led to the development of more derived birds, is still an area that is being worked on.

Ancient representatives

The exact appearance of the modern birds (that is, any species that is part of a lineage with members still alive today) is also a subject of great interest, probably because of the extreme popularity of birds in general. Perhaps the oldest known so far is a bird called *Vegavis*, which is from Antarctica and dates back to about 67 million years ago, and so very close to the extinction event. *Vegavis* was mentioned in an earlier chapter, as one of the specimens of this animal is preserved with part of its trachea and gives us an idea of the kinds of sounds it could make. One of the inferred noises is a goose-like honk, and indeed *Vegavis* is considered to be part of the lineage that gave rise to modern ducks and geese and other waterfowl.

As ever, a fossil of this kind gives us only a minimum date for the origins of the modern birds, since there are likely to be older fossils than this and indeed we would expect them to be out there. The lineage of ducks is known to be an ancient one and its nearest relative is the branch of birds that gives us chickens, turkeys, pheasants and other game birds, giving rise to the possibility of Cretaceous chickens (well, chicken-like ancestors at least).

But this pair of groups, which go by the tongue-twisting name of Galloanseriformes, are not considered the first branching group of modern birds, which is a clade called the palaeognaths and consists of the odd little flying tinamous and the much more well-known ratites – ostriches, emus, cassowaries, rheas and the kiwis. These latter birds, of course, are all flightless, though this loss of flight evolved separately numerous times in this group.

The important point is that they must have branched off before the origins of the Galloanseriformes and thus are probably rather older than *Vegavis*, but quite how far back they go isn't known. This is the kind of problem that's unlikely to be solved without additional fossils, but it's one that adds a very intriguing note to looking at the birds that are with us now, and the question of how far back they do go.

One area that has been the subject of much debate is to what degree the evolution of birds was influenced by the presence of the pterosaurs. These flying reptiles are themselves very close relatives of the dinosaurs and had been around for perhaps 70 or so million years when the first birds became powered fliers. The presence of two flying vertebrates at one level should not be a surprise, since today we have both bats and birds in huge numbers of species and individuals, although the case here is rather different.

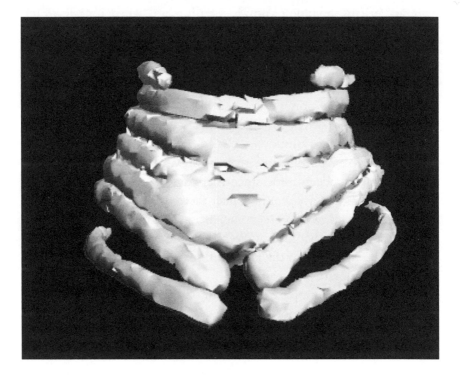

Digital scan of the fossilised 'voice box' of the Cretaceous
Antarctic bird *Vegavis*. Image by Julia Clarke.

There are various diurnal bats and nocturnal birds, but there's a fairly obvious split in their activity, which represents something of a separation in their ecology. However, both early birds and various pterosaurs were mostly diurnal and the assumption is that at some level they must have clashed. One major lineage of early pterosaurs did die off and another radiated greatly not too long after the origin of the birds, but quite what level of interaction there was between these various factions is very uncertain.

Getting into the later Cretaceous, many pterosaur groups became extinct and others reduced greatly in diversity, and while they persisted until the end of the Mesozoic, they were nothing like they had been before. It seems improbable that the spread and diversification of another group of flying animals could have had no effect on pterosaur diversity, but equally pterosaurs lived alongside birds for perhaps a hundred million years, and it was tens of millions before any real dip in pterosaur diversity really kicked in.

In addition, pterosaurs clearly occupied some niches that the Mesozoic birds never occupied, most notably the huge marine soarers like *Pteranodon* with a wingspan of 5–9 metres. Also, the anatomical organisation of pterosaurs meant they were probably not very good in the trees, one area where birds would excel, and so at least some groups of these should have been pretty separate from one another and had little, or even no competition. The very limited fossil record of pterosaurs, especially in the middle of the Cretaceous when this apparent dip in diversity occurs, makes this very hard to establish, and various papers have argued both for and against a major influence of one or the other.

Happily, pterosaur research is enjoying a real boom at the moment, with a dramatic increase in the number of researchers and in the number of fossils being described, so this is an area that is likely to get some better resolution in the coming years.

Survival

The pterosaurs and the non-avian dinosaurs were all lost at the K–Pg extinction event, but the birds made it through. The popular conception of this is that the birds were survivors and made it over the line relatively unscathed, but the truth is that very large numbers of bird lineages also went extinct. This is perhaps not surprising given the extent and severity of the extinction, but in many ways birds seemed well equipped to escape.

They were generally small and numerous, meaning that they had an increased number of individuals and species that might get lucky and that wouldn't require much in the way of resources in the aftermath, but, most importantly, they could fly. With such a global catastrophe, being able to find, and then move to, a relatively unscathed location would be very important, and flight should have given birds a huge advantage over the vast majority of animals in being able to do this.

It does come as quite a surprise, therefore, to learn that the birds that made it across the K–Pg boundary were the ground dwellers.

These were not flightless birds, but were species that spent the majority of their time on the ground and, by extension, were not especially good at covering long distances, and so don't fit the obvious profile of extinction-event survivors. What appears to have been the major issue is that trees did especially badly during the extinction and so any birds that favoured arboreal habitats (that is to say, most of them) suffered, as their habitats were destroyed and they too perished.

Inevitably, there were still some exceptions. Several aquatic lineages of birds, including the flightless grebe-like swimmers the Hesperornithes, also went extinct. While they couldn't fly, being able to move freely around coasts or even across seas should have given them a good shot at surviving; they were not, of course, tied to the trees at all.

In short, while the main reason for the loss of most birds across the K–Pg seems to be in hand, the details are still there to be worked out and there is much to do here and elsewhere in our understanding of the evolution and diversification of the birds during the Mesozoic. Despite the plethora of new finds, these have been concentrated in areas of exceptional preservation and so the fossil record for most of the Cretaceous remains relatively scant and thus limits our understandings.

However, with the huge amount of attention that early birds draw, research here is especially intensive and the massive strides made in our understanding of bird origins in the last twenty years is only likely to increase still further and faster in the coming decade. Many of the remaining questions and uncertainties here are likely to be resolved sooner rather than later.

16

Research and Communication

WHEN I WAS an undergraduate student and I wanted to find a paper on a subject, I would go to the departmental library and go through a series of fat volumes that were published each year, which had huge indexes of keywords associated with papers. If I were interested in a paper on lion behaviour, I'd look through lists of 'lion', 'felid', 'behaviour', 'ecology', 'carnivore', and perhaps a few others, and get checklists of papers that had referenced these words.*

Based on the titles of these papers, I'd get an idea of which might be useful and then I'd head into the library proper to find the respective journals on the shelves and the issue that contained the paper. Often we would not have it, because we didn't stock every journal out there, or we did but not far back enough to get that specific paper. I might have to get a loan from another university, or write to the authors and get a copy sent in the post. Titles and keywords are not always that informative, though, and often when I finally got my hands on something, it would turn out not to have anything like the information I was looking for.

On top of this, any recent papers wouldn't have been in the volume of keywords (which came out annually), meaning a manual check of each of the recent issues of a few dozen journals in racks at the library entrance. It was not uncommon to spend half a day like this chasing paper trails to get two or three useful papers out of the exercise, and

---

* There was also an early online system for searching for papers at this point, but it wasn't very good and most universities didn't have it. The process I describe was something I did more than once and would have been near universal for students or academics only a few years earlier (in the early 1990s or before).

it was a less than thorough search of all the available information and papers that were actually out there.

By the time I started my PhD in the early 2000s, it was common to be able to email authors, colleagues or even libraries and get hold of scans of papers, or even electronic copies of papers both old and new. Searching for papers online (even if you then had to look for the hard copy in the library) was common, and you could at least now read the first page of a paper and not simply the title to see if it was relevant. Some journals were even online and it was possible to get hold of and read the papers without leaving your desk.

Academics started to move from having vast stockpiles of printed and photocopied papers in their offices to hard-drives on the computer with the relevant literature on it, and it became possible to copy a whole decade's worth of information in a few hours, rather than taking a pile to the library and taking notes from a paper you might never see again, or photocopying it a page at a time.

By now, of course, almost everything is digital and it's possible to access sometimes centuries-worth of publications from a single journal in a couple of clicks, and there are multiple online databases and search engines that allow you to go through thousands of journals and articles and skim them for relevance, or pull entire pools of related papers. In barely twenty years, we went from perhaps taking a day to get half a dozen relevant papers to getting hold of most of the available work on a single subject in a few hours. You still have to read it, but the labour saved is almost unimaginable.

I go through this not to complain that 'kids today don't know how tough it was', but to show how vast and dramatic and unexpected this change has been. I'm not sure anyone really anticipated this shift, or if they did, how quickly it would come about. The change in accessibility to knowledge and the culture of now being able, and expected, to have dug much more thoroughly into the literature has clearly improved the quality and drive of scientific research, but it's also a quite unexpected benefit. There is often talk about what the next generation of research techniques might be that will become available to researchers (and I will do it here too for palaeontology), but aspects of change like this are quite unpredictable and can have a greater benefit to science as whole than the simple increase in computing power, or the next type of scanner becoming accessible.

Tech takeover

Similar to the point above (and allowing me to indulge in a bit more anti-nostalgia for how things used to be worse in my day), I was initially trained to do evolutionary relationship analyses by hand, as, while it was long-winded, you could do larger analyses this way than could the computers of the late 1990s, as they could barely handle ten species.

By 2002, as I started my PhD, everything was done on computers and the analysis for my final piece of work had a huge fifty-plus species in it, though the analysis took nearly a week of computer run time (not that I had an especially powerful machine admittedly). By the time I finished in 2005, new software had cut that processing time down to about two seconds.

This was always primarily a raw power issue, and with bigger and better computers running newer and faster software, things got better quickly, and we can be pretty sure that computing power will continue to increase and that ever greater analyses of all kinds with greater processor power will be possible.

Other aspects of technological drives are far less obvious, or their applications may be unexpected or unavailable. Medicine clearly drove things like X-rays and CT scanners, which are now common-place for investigating fossils. Palaeontologists probably saw the poten-tial applications of these very early on, but emerging technology is rare and expensive to use, and palaeontology is not a field flush with money. It would also take considerable experimentation to get things set up to where they would work. Some things are surprisingly resist-ant to being assessed and various flattened specimens have proven very hard to scan or penetrate with any success, as the bones are simply too close in density to the entombing rocks for the various scans to be able to distinguish between them.

So while there are incredible scans into the skulls of huge tyranno-saurs, it may take yet greater resolution to peer into some quite small early birds and other smaller dinosaurs. I must confess to not being up to speed on the latest medical scanning and imaging technologies, or what might be coming in the next generation of data visualisation and analyses associated with such things, but it would be strange indeed if these were not set to improve or new ones appear soon, and stranger still if they were not exploited by future palaeontologists.

One other major factor that remains unknown is how things will play out in terms of the ability of researchers to carry out their work. Most palaeontologists are based in museums and universities, but recent years have seen shortages of funds in the former and far-reaching changes in the latter, with more likely to come. Palaeontologists remain few in number, and while dinosaurs themselves are popular, dinosaur researchers remain a smaller group within that number. When institutes are looking to make savings, cutting jobs that affect only a small discrete pool of people are often high on the list, so palae-ontologists are perhaps more at risk of losing their support than others. With funds tight for students as well, a degree that offers less security can also be less favourable, and as a result the people who support university academics can dry up.

Moreover, the sources of research funds can change or be redirected. Major grant-giving bodies have different priorities to the various fields that they govern, and while it may be entirely sensible that money is concentrated into conservation and climate-change research right now, that will undoubtedly leave other fields behind. This can drive research enormously, as scientists will push their research into areas that are looked upon more favourably and likely to increase their chances of continuing their work. That means palaeontology will still be done, but areas like taxonomy might be massively reduced while other areas such as extinction are focused upon.

That in turn means a fundamental shift in the patterns of research, and promising lines of enquiry, or new areas ready to be explored, will have to wait until later, or at least take second place. This is not true globally, and China has invested massively in science generally and palaeontology specifically in recent years, but naturally each country and funding body will prioritise different things, meaning shifts that are not necessarily directed by the researchers themselves can come along and move the direction of entire fields.

Academics themselves will also drive research directions and these are potentially even harder to predict. New fields arise (either by becoming new areas of research or rising to prominence), such as the relatively recent explosion of evolutionary-developmental biology, called evo-devo for short. This branch of biology combines palaeontological data and that from broad evolutionary studies of lineages with information from the development of zygotes and embryos as they develop, to look at deep changes and relationships between various groups and how they have evolved. Evo-devo has had massive attention from biologists and palaeontologists, as it helps unite the fields and has provided some answers to long-standing questions, as well as yielding new ideas and insights ripe for further testing.

However, this is now an established branch of science and there are entire labs and research teams committed to it, with some scientists likely to devote their careers to this one area. That's no bad thing, but again, if evo-devo is attracting palaeontologists to its banner, they will by definition not be working on other areas, and so will act to redirect years or decades of research. There are only so many researchers and so many projects they can work on, and while science will continue to move forwards, the rates at which some areas advance will vary enormously. It's all but impossible to tell which will be next and the implications this will have on what we discover in the coming years.

Future finds

Unlike many branches of science, palaeontology is often limited by what things are available to study. We cannot work on the feeding ecology of a dinosaur that doesn't have a known skull, or look at the likely locomotion of one that lacks legs. Burrowing or swimming dinosaurs were all but hypothetical until the discovery of species that seemed well adapted for these ecological niches and, despite the presence of *Archaeopteryx*, there was little to say about the feathers in dinosaurs, and certainly not deep into the theropods, until the discovery of various other feathered forms in the late 1990s.

This example above all others is one where quirks of history could have dramatically changed dinosaur research and in particular the discovery of the dinosaurian origins of birds. Although this had been hypothesised relatively early on in the history of dinosaur research, as an idea it fell by the wayside through lack of evidence until a resurgence in the 1970s, and then vindication with new feathered species in the 1990s and 2000s. We now know that there were feathered dinosaurs present in the Solnhofen limestones that yielded *Archaeopteryx*, and so they potentially could have been found as far back as the late 1700s when these stones were first exploited for their fossils, and certainly in the mid-1800s when the 'first bird' was uncovered.

Similarly, the great fossil beds of Alberta have only yielded feathered fossils in the last few years, but dinosaurs have been excavated there for a century now, and finally, the enormous fossil beds of Liaoning in China that have produced the vast majority of the feathered dinosaurs were worked on in the 1930s by Japanese geologists. Had any of these locations produced a feathered dinosaur (or even many), then the bird – dinosaur link might have been established much earlier.

Were that the case, it's hard to imagine that the amount of disagreement over the activity levels of dinosaurs or the arguments over their posture could have persisted to the degree they did, and given the apparent endless fascination for birds, this could have driven a great deal of interest in the dinosaurs in what was otherwise a fallow period in the first half of the twentieth century.

It is hard to imagine that an equivalent discovery could occur right now that might dramatically change our understanding of the dinosaurs. Even the apparent fluidity of the early divergence is no enormous change to the family tree, and the general evidence that says all dinosaurs are related to one another and are archosaurs is essentially overwhelming.

However, as shown across this book, there are significant gaps in our knowledge and any of these might be filled with a new find. For example, a new early dromaeosaur or troodontid would make a huge difference to resolving their exact relationships with each other and the birds, and could tell us a lot about flight in these groups. There are fossil beds in parts of Russia that are starting to yield early ornithischians with unusual filaments that may be linked to early feather evolution, and there may be many more like these in the area. They are likely to be very productive in the future, but it's hard not to yearn for them to have turned up twenty years ago, or for another set of similar beds similar.

We still have very few fossils of large animals bearing feathers and generally few mummies or animals that are exceptionally preserved, and it's entirely possible that new locations and formations will be uncovered that will produce a huge number of new and important fossils and begin to help resolve some outstanding questions.

Already we are beginning to find bits of dinosaurs (and especially birds) preserved in amber coming from the Cretaceous deposits of Myanmar. We're never going to find something large like a spinosaur or a hadrosaur in amber, but bits of dinosaur skin, feathers, tiny individuals, or things such as fingers and tail tips could all be preserved. This would potentially give us new insights into muscles, nerves, blood vessels, air sac structure and other tissues or organs. Coupled with the increasing interest in, and data from, specimens that appear to preserve original organic molecules, and this is likely to be a major area that can leap forwards again with new discoveries, especially if new sites that yield many fossils are found.

Much of what is changing our ability to interpret dinosaurs in ever more detail is based on fields like evo-devo and an ever increasing knowledge of the biology and evolution of living species that we can use for comparisons. It was only in the last few years that circular breathing was demonstrated in crocodiles and lizards, suggesting that this evolved before the origins of air sacs and that such a system of moving air through the lungs might be universal for dinosaurs. But this kind of research is not usually in the hands of dinosaur researchers and we are somewhat at the whim of what our colleagues in biology are working on, and what they may discover that will become applicable to us in the future.

There is also an increasing integration between these areas where they used to be very separate domains, and this itself is likely to drive a much better understanding of dinosaurs. Biologists are increasingly seeing the data that palaeontology can bring giving evolutionary contexts and information derived from changes over tens of millions of years, which are impossible to get from living groups alone, while palaeontologists benefit from the depth of information and under-standing from studies of living animals. It is a trend that I certainly hope will continue to the benefit of all (though with the caveats outlined above as to the future of research), but as to whether or not this does continue, and if so which branches of biology will be most involved, it's impossible to say.

Communication

Carrying out research to learn new things is obviously a major part of any field of science, but the results generated and conclusions reached also need to be communicated effectively both to academics and members of the public. Again, the internet and other technologies have driven this forward at an incredible rate and utterly transformed how scientists communicate on a daily basis. Email means that you can contact a colleague anywhere in the world and get an instant reply, rather than sending a letter and hoping it arrives in a week or so, and it's possible to have video chats and share papers or images quickly and easily.

That transforms the abilities of people to communicate, but also more importantly to collaborate – if you have found an unusual bone and you think it might belong to a group with which you are not familiar, send a few photos to the world expert to advise you. Neither of you needs to trek halfway round the world to meet up or share data, and you can easily speak to as many people as you want and canvass opinions of a number of different people, and then come together digitally to write scientific papers. I've contributed or co-authored papers with a number of people that I have never met in person and this is increasingly common.

Other related technologies are also facilitating the exchange of information and in particular those related to 3D imaging in various forms. CT scans and other virtual 3D models have been around for a while, but they were expensive to produce, the resolution tended to be low, and not many people had the software to be able to view them effectively. Now those barriers have been removed and such models are increasingly being used. One technique of particular note is called photogrammetry, where multiple digital photos are combined to make a model of an object.

Photogrammetry is wonderful for a number of reasons, but especially because it requires little skill or expertise. A bunch of photos from different angles is all that is needed to make a half-decent model of even a complete dinosaur skeleton. Obviously taking more photos and ones that are well shot and in focus and cover key details is preferable, but the software can easily cope with some poor lighting and out-of-focus images if others are better, and some remarkable details can emerge from very little.

It's also trivial to take photos of something that you can't actually access (like a mounted skeleton in a museum, especially if it's behind glass), but with a generated model you can then take measurements in places or of features you could not otherwise reach, or a specimen you could not physically move. Some lost or damaged fossils have even been restored with these methods, scanning into the computer all manner of archival photographs to remake them and allow them to be seen again.

With such 3D models also comes the opportunity to 3D print them and make physical copies. There is something wonderfully tactile about such models, which an image on screen can never quite replicate, but it also gives the opportunity to scale up or down and so large bones become much easier to manipulate and small ones easier to see. You can also begin to repair specimens digitally, making models more accurate for display and restoring them to their former glories. You can even start to 'average' multiple specimens to generate a best understanding of a typical individual of a species and then compare these averages for a more accurate impression of evolutionary changes.

Having such digital physical models is great for academics but also for the public, as they can see or handle versions of things much more easily that might otherwise be out of reach, either because they are in a museum many miles away, or because they are too fragile or valuable to be displayed, let alone to be handled. Museums can also use these digital physical models to create ever more elaborate and interactive exhibitions, and it is increasingly common to see exhibits that use VR or augmented reality displays. This is only likely to increase.

Digitally restored 3D scans of the skull of the small tyrannosaur
Dilong. The differing shades represent different stresses on the
skull when loaded, as if it was biting something to see how strong
it was. Such work was impossible before modern computing.
Image courtesy of Evan Johnson-Ransom and Eric Snively.

However, it's hard to see where this side of things may go next, if
only because museum budgets are often limited and it can cost
millions to move and remount a single dinosaur skeleton, which
may be the priority if it's stuck in a 1960s tail-dragging pose, rather
than generate a new VR model of it. Hopefully this will expand
further and become more integrated with the fossils themselves and
bring in a new level of communication of science to a general
audience.

In a similar vein, modern technology has massively enhanced
communications between researchers and the public. For as long as
there has been science, there has been a dialogue between researchers and those outside the world of academe, in particular through
public talks. Some years ago, staff at one UK university were asked
to shift their teaching style dramatically to fit the new generation of
students, and to make as much as possible linked to online materials
and videos of lectures, and so on. The response from the faculty was
along the lines that lecturing an audience had been appropriate and

even perfected over the last two and half thousand years and there was no need to go and change it now.

Times have moved on, of course, but the lecture remains a primary method of communicating from academic to audience, be they students or the public. Appearances in the media, from newspapers and magazines to TV and radio shows, helped too, but the fundamentals were always direct communication in person. Palaeontologists still engage in all of these aspects, but now they also write blogs, post stories and photos online, join message boards and answer questions, join in online forums, and deliver lectures online, through podcasts and all manner of other methods. They can reach out in a myriad of new ways and, especially over distances, with a rapidity that was previously unimaginable.

It is also very much a two-way process. I have a number of friends who are very proud of the written letters they received as responses to enquiries they had sent to various palaeontologists when they were growing up. They were delighted to hear back from their idols, or simply people who could help them out, but it is notable how rare an event that was. Finding the address of a researcher was not necessarily that simple and taking the trouble to write to them meant this was not very common.

These days it's a rare week that I don't get an email or notification from a member of the public about some aspect of palaeontology, and I can get multiple ones in a single day. That is mostly an enormous boon to keen communicators of science and of course to the public as well; and again, all the benefits above of being able to share images or link to existing materials online are there too, making it far more effective. I should point out that it's not all roses, however; it is easier generally for people to find you, but that does mean everyone, and so it also increases the volume of communication from creationists, cranks and other people, who can take up a lot of time without you necessarily being able to help.

It's also hard not to feel bad when you are telling yet another small child that the rock they found on the beach is not a new *Tyrannosaurus*, or you have to tell people you simply don't have the time to answer their list of fifty burning questions they have about dinosaur claws. On the whole, it's a massive positive for all concerned, but I don't look forward to the day when someone takes real exception to a point made by an academic that destroys their pet theory.*

There has always been a place for non–academics ('amateur' can imply they are not professional in their work, which is not usually the case) to get involved in research, but that has also grown rapidly. With papers and researchers more accessible, fossil finds by the public are more likely to reach a museum and be identified correctly. People offer their services and I've been delighted to have collaborated with and even published papers with those, who have little or no formal background in science but have used their skills to help make a real contribution in the sciences.

Palaeontologists are engaging with this ever more and we now see researchers crowdsourcing data, and even crowdfunded excavations, preparation and publication of new finds. All of this has come about through increased and improved communications between academics and the public, and again this is an area that is only likely to increase further. Researchers have already been using drones to take photos of fossil sites and to hunt for specimens, and it can't be too long before these become live feeds, with people watching palaeontologists dig online, or even helping to ID things from the air as we go.

It is notoriously hard to anticipate where technology will go next (as all those magazines predicting flying cars, jetpacks and moon bases will attest) and forecasting the intersection between technology, palaeontology and the public might be beyond the prognosticating powers of anyone who would care to try. Still, short of a new technological dark age, it is hard not to expect this to continue to leap forwards. I'd certainly hope, though, that were will always remain a place for a good old-fashioned (even Socratean) lecture and discussion with a real audience.

---

* The internet does massively embolden cranks, who can put up enormous essays and details about their pet 'theories'. It allows them to email or tag large numbers of researchers repeatedly. Many are simply misguided or misinformed, but don't want to hear that they may be wrong; sadly some are extremely antagonistic and downright nasty. The ideas themselves range from the obviously wrong to the downright bizarre, and no amount of evidence or fundamental flaws in their arguments will convince them otherwise. A friend of mine has repeatedly suggested I write a book on palaeontology cranks, and it may yet happen one day.

17

Coda

WHILE THIS BOOK attempts to show what we don't know (yet) about dinosaurs, it does hopefully also demonstrate just how much we do know. At the time that the word 'dinosaur' was coined by Richard Owen in 1842, we knew of just three species of dinosaurs, represented by a handful of rather incomplete specimens that all came from southern England.

Astute observers might have realised that there would be more to find and that the Age of Reptiles probably featured plenty of large terrestrial saurians, but it is hard to imagine that they would have been able to predict the vast swathe of animals that would be uncovered in the coming decades and continue to be found nearly 200 years later. Dinosaurs are now represented by huge numbers of specimens and even if most species are known from frustratingly scant remains, we have dozens of taxa that are represented from large numbers of exquisitely preserved fossils and all manner of soft tissues, in addition to bones and teeth.

Science as a whole has developed enormously in this time and palaeontology has gone from not even being a recognised field, to being a lesser branch of geology, to becoming a major science in its own right with various branches itself. Palaeontologists now have a good handle on the origins of dinosaurs, the major groups that evolved, how long they lasted, and their relationships to one another. We understand the various causes of the extinction of the dinosaurs, and probably what allowed their surviving membership in birds to get through that difficult time and expand and diversify. More and more we understand the patterns of evolution that the dinosaurs took and what that means for their history on this planet.

Fundamentals of research in dinosaurs' anatomy, physiology and biomechanics have accelerated hugely and so we understand how they were built (even sometimes at the cellular or biochemical level), so we can flesh out the skeletons and see how muscles, tendon and bone combined to allow them to move. Even more detailed have become our reconstructions of their external appearances with the discoveries of fossil scales, feathers, claws, and onto patterns and colours. New research has unveiled how they lived as animals, where they lived, what the environment was like with its predators, prey and competitors, as well as what these animals fed on, their migration patterns, how they behaved, and how they reproduced and grew to found the next generation.

These are not trivial issues. It is true that a large number of such ideas would have eventually come to the fore through the discovery of relevant specimens – you don't have to be brilliant to work out that pine cones inside a hadrosaur are likely there because it ate them – but this is not the case for most.

Even with apparently simple areas such as diet, researchers need to generate and then pull together data on tooth and jaw shape, how muscles attached to them, the structure of their beak, the mechanics of the jaws in operation, microwear on the teeth, gut contents, and known plants from ancient pollen samples, determining if such pine cones were an occasional or habitual part of the diet, if the local environment would support these plants in large numbers, what competitors might also feed on them, what this means for the local climate or the structure of the ecosystem and more. All of this might need to be analysed to give a firm answer beyond 'some pine cones at least occasionally'. That can take years of work and relies on such a diverse skill set and different types of data and analyses that it would be impossible to be completed by a single person and belies the breadth and depth of the research required. This is why combining all these disparate branches and lines of evidence really can provide such a huge increase in our understanding of the dinosaurs and their world.

Just how far we have come is therefore impressive, but it pales against what we are yet to learn or can never know. Take ourselves, for example: humanity. Understandably we have always had a particular fascination with our own biology, our evolutionary origins and history, and how we came to be.* Humanity has done more research on itself than on any other species, which has a history that extends further back than perhaps any other science and shows no signs of slowing down. The medical literature alone accounts for tens of thousands of papers annually and has done so for decades, and we are still learning details of fundamentals like our anatomy (something you might think would have been resolved decades, if not a century ago).

There is so much to learn about our own species and we don't appear to be near the point of having worked out most of it. Even if much of what we still have to uncover is devoted to biochemical pathways and genetic data that can never be accessed and resolved for the dinosaurs, there are more researchers working on human health alone in one institute in the UK than there are academics at all the world's universities and museums working on dinosaurs. Given the colossal head start that human research enjoys, and the huge advantages in funding and number of researchers, it should be a wonder that we know as much as we do about ancient reptiles.

The gaps are clearly very large, but the big picture is there. I think a great analogy here is one of a giant and detailed picture puzzle, with the final result being all that can be known about dinosaurs. We know that we have lots of bits missing and can never find them, and plenty of the pieces we have are themselves incomplete and don't fit together as cleanly as they should, and some could go in multiple places. But all of that said, we don't need that much to get a feel for what the final picture would look like.

Some large areas are about as complete as they can be for us to see what that part looks like, and make solid predictions about what the missing parts will look like if we find them. Large resolved bits of the puzzle fit together, so we can see how different parts are related. Some of these links are very strong and others weak, and may change later if we get some new bits, but we can recognise that these connections are weak and if things do go somewhere else, we can probably see now where that might be.

---

* Leaving aside the issues over our intelligence, which is especially complicated and rather takes away from the parallels that I'd like to draw with other animals, as humans are clearly a huge outlier when it comes to mental capacity and the intricacies of our intelligence and social interactions.

Occasionally some big chunks will move, but the overall pattern within them is good and it won't change the rest of our knowledge. In short, the picture might have more bits missing than are put together, but we do know whether it's going to be a picture of a country scene or a cityscape, and we have plenty of details, too. To extend this metaphor a fraction further – we don't know what bits of the puzzle we may yet draw from the box, though we have some idea of where they may go, and we do know some ways of getting more out of what we have.

Going forwards

What then is most likely to break next? Which areas are likely to extend greatly because of a technological breakthrough, a new level of interest, or new understanding? These are naturally difficult predictions to make, but it would be remiss of me not to attempt to do so. Before we look at each of these three drivers, there is an additional one that is impossible to predict: what fossil finds may drive this? Areas such as the Chinese beds of feathered dinosaurs, or the massively productive beds from the Morrison of North America and the bizarre island faunas of central Europe have all helped in their ways to bring about major breakthroughs in our understanding and knowledge of dinosaurs.

While we have a great idea of the geology of the planet, that does not mean that major areas are not under-explored. They are rife with possibilities. New fossil localities are being found all the time, and those that have been overlooked as being of little interest or unproductive may yet prove to be treasure troves of huge amounts of new and exciting finds, which can kick-start new fields and interest. These are surely coming and I would suspect that sooner or later something in East Africa, or Australia in particular, is likely to become the next big thing.

Turning to the three areas that I think will drive new discoveries, first there is technology. Here the obvious choice would be molecular palaeontology and the increasing possibilities from fossil protein fragments and others ancient molecules. However, it is not clear how far this field can go, as it is too reliant on only the select fossil localities and specimens that can yield any data. With that limitation, there remains the possibility that there will be too few comparisons that can be made between specimens and species, and while what it will find will be incredibly interesting, it will struggle to become a major part of dinosaur palaeontology.

Instead then, I would go with the growing area of biomechanics. Ever increasing computing power and decreasing costs of things like 3D printing are allowing a far better understanding of different parts of dinosaurs and how they fit and work together. We are getting better and better able to explore how heavy they were, how fast, how agile, what this would mean for their posture, ecology and behaviour; and few things influence the biology of an animal like its size. This has so many implications that I'd expect the increase of work in this area, prompted by these technological facilitations, to be a major advance in the coming years.

Second, there is currently a huge new (or renewed) interest in the behaviours of dinosaurs. While we have always chipped away at this area in basic terms, we are only now beginning to delve into complexities and away from raw speculation. With our abilities to determine colours and patterns of dinosaurs, as well as details such as differences in feeding and new data on groups of dinosaurs and their possible intelligence, there is the beginnings of a surge in this area.

It is tentative right now, but I think that with the new data coming in these areas, more and more attention will turn to how these animals lived and interacted with one another, especially communications within species. Hopefully this will blossom soon into a new level of understanding of the lives of dinosaurs and what this meant for their evolution and diversity.

Finally, the improvements and interest to come from these new levels of understanding. Here I will go with ecology – already, of course, a huge and established field of biology in its own right, it is developing rapidly with the integration of molecular data. Ecologists have an ever greater understanding of how modern ecosystems function, how species interact, how energy flows within and between systems, and how populations of one species affect another. As this knowledge improves rapidly, so palaeontologists are able to apply those newly verified rules and ideas to ancient ecosystems and learn more about the world of the dinosaurs. Palaeoecology is already developing, but a massively improved basis for interpretation can only help it to accelerate.

Such predictions are always fraught with uncertainty and even those best placed to judge can be completely wrong (i.e., I have only so much faith in my own predictions and they are at least partly driven by hope as much as expectation). However, whichever field benefits most from whichever development, it is certain that our knowledge base of facts, our understanding through interpretation and our extrapolation through theory will increase. The gap between what we do know about dinosaurs and what we do not may be large, but it is closing and rapidly. Long may that continue.

This spectacular tyrannosaur skeleton recently went
to the North Carolina Museum of Natural History.
What secrets will be revealed when it has been prepared
and studied? Image courtesy of Lindsay Zanno.

References

There are tens of thousands of scientific papers on dinosaurs as well as all kinds of books, collections of research, conference proceedings, abstracts and postgraduate theses and dissertations, and the list grows by thousands more each year. For any of the chapters in this book I could have cited many hundreds of different sources, both historical and recent, many of which would not necessarily agree with one another. I did not want to bog down the text of the book with endless references or to ignore the huge contributions of so many academics to the fields of dinosaur research, so in an attempt to find a middle ground, below are some selected works for each chapter. The intention was to try and include those that covered a wide range of topics or at least dealt with a major point raised in the chapter. Of course the one area these papers don't tend to cover is the core subject of this book – what we don't know about dinosaurs. Perhaps unsurprisingly, most scholarly works deal with the discovery of new knowledge rather than the unfilled expanse beyond.

Academic papers are not always easy to access or necessarily easy to read, but for those who would like to engage further with the scientific literature, here are some key places to start.

Introduction

Godefroit P. ed., 2012. *Bernissart dinosaurs and Early Cretaceous terrestrial ecosystems*. Indiana University Press.

Mantell, G.A., 1825. VIII. Notice on the *Iguanodon*, a newly discovered fossil reptile, from the sandstone of Tilgate, in Sussex. By Gideon Mantell, FLS and MGS Fellow of the College of Surgeons, &c. In a letter to Davies

Gilbert, Esq. MPVPRS &c. &c. &c. Communicated by D. Gilbert, Esq. *Philosophical Transactions of the Royal Society of London*, 115, pp. 179-186.

Chapter 1: Extinction

Alvarez, L.W., Alvarez, W., Asaro, F. and Michel, H.V., 1980. 'Extraterrestrial cause for the Cretaceous-Tertiary extinction'. *Science*, 208(4448), pp. 1095-1108.

Benton, M.J., 1989. 'Scientific methodologies in collision: the history of the study of the extinction of the dinosaurs'. *Evolutionary Biology*, 24, pp. 371-400.

Keller, G., Sahni, A. and Bajpai, S., 2009. 'Deccan volcanism, the KT mass extinction and dinosaurs'. *Journal of biosciences*, 34(5), pp. 709-728.

Sakamoto, M., Benton, M.J. and Venditti, C., 2016. 'Dinosaurs in decline tens of millions of years before their final extinction'. *Proceedings of the National Academy of Sciences*, 113(18), pp. 5036-5040.

Chapter 2: Origins and Relationships

Baron, M.G., Norman, D.B. and Barrett, P.M., 2017. 'A new hypothesis of dinosaur relationships and early dinosaur evolution'. *Nature*, 543(7646), p. 501.

Brusatte, S.L., Nesbitt, S.J., Irmis, R.B., Butler, R.J., Benton, M.J. and Norell, M.A., 2010. 'The origin and early radiation of dinosaurs'. *Earth-Science Reviews*, 101(1-2), pp. 68-100.

Langer, M.C., Ezcurra, M.D., Bittencourt, J.S. and Novas, F.E., 2010. 'The origin and early evolution of dinosaurs'. *Biological Reviews*, 85(1), pp. 55-110.

Lloyd, G.T., Davis, K.E., Pisani, D., Tarver, J.E., Ruta, M., Sakamoto, M., Hone, D.W., Jennings, R. and Benton, M.J., 2008. 'Dinosaurs and the Cretaceous terrestrial revolution'. *Proceedings of the Royal Society of London B: Biological Sciences*, 275(1650), pp. 2483-2490.

Nesbitt, S.J., 2011. 'The early evolution of archosaurs: relationships and the origin of major clades'. *Bulletin of the American Museum of Natural History*, pp. 1-292.

Chapter 3: Preservation

Arbour, V.M. and Currie, P.J., 2012. 'Analyzing taphonomic deformation of ankylosaur skulls using retrodeformation and finite element analysis'. *PloS one*, 7(6), p. e39323.

Dal Sasso, C. and Signore, M., 1998. 'Exceptional soft-tissue preservation in a theropod dinosaur from Italy'. *Nature*, 392(6674), p. 383.

Eberth, D.A., Xing, X.U. and Clark, J.M., 2010. 'Dinosaur death pits from the Jurassic of China'. *Palaios*, 25(2), pp. 112-125.

Falkingham, P.L., Bates, K.T., Margetts, L. and Manning, P.L., 2011. 'The "Goldilocks" effect: preservation bias in vertebrate track assemblages'. *Journal of the Royal Society Interface*, 8(61), pp. 1142-1154.

Lauters, P., Bolotsky, Y.L., Van Itterbeeck, J. and Godefroit, P., 2008. 'Taphonomy and age profile of a latest Cretaceous dinosaur bone bed in far eastern Russia'. *Palaios*, 23(3), pp. 153-162.

Chapter 4: Diversity

Barrett, P.M., McGowan, A.J. and Page, V., 2009. 'Dinosaur diversity and the rock record'. *Proceedings of the Royal Society of London B: Biological Sciences*, 276(1667), pp. 2667-2674.

Brusatte, S.L., Benton, M.J., Ruta, M. and Lloyd, G.T., 2008. 'The first 50 Myr of dinosaur evolution: macroevolutionary pattern and morphological disparity'. *Biology Letters*, 4(6), pp. 733-736.

Fastovsky, D.E., Huang, Y., Hsu, J., Martin-McNaughton, J., Sheehan, P.M. and Weishampel, D.B., 2004. 'Shape of Mesozoic dinosaur richness'. *Geology*, 32(10), pp. 877-880.

Wang, S.C. and Dodson, P., 2006. 'Estimating the diversity of dinosaurs'. *Proceedings of the National Academy of Sciences*, 103(37), pp. 13601-13605.

Chapter 5: Evolutionary Patterns

Maidment, S.C., Henderson, D.M. and Barrett, P.M., 2014. 'What drove reversions to quadrupedality in ornithischian dinosaurs? Testing hypotheses using centre of mass modelling. *Naturwissenschaften*, 101(11), pp. 989-1001.

O'Gorman, E.J. and Hone, D.W., 2012. 'Body size distribution of the dinosaurs'. *PLoS One*, 7(12), p. e51925.

Organ, C.L., Shedlock, A.M., Meade, A., Pagel, M. and Edwards, S.V., 2007. Origin of avian genome size and structure in non-avian dinosaurs. *Nature*, 446(7132), p. 180.

Puttick, M.N., Thomas, G.H. and Benton, M.J., 2014. High rates of evolution preceded the origin of birds. *Evolution*, 68(5), pp. 1497-1510.

Upchurch, P., Hunn, C.A. and Norman, D.B., 2002. An analysis of dinosaurian biogeography: evidence for the existence of vicariance and dispersal patterns caused by geological events. *Proceedings of the Royal Society of London B: Biological Sciences*, 269(1491), pp. 613-621.

Chapter 6: Habitats and Environments

Fricke, H.C., Rogers, R.R. and Gates, T.A., 2009. 'Hadrosaurid migration: inferences based on stable isotope comparisons among Late Cretaceous dinosaur localities'. *Paleobiology*, 35(2), pp. 270-288.

Mallon, J.C. and Evans, D.C., 2014. 'Taphonomy and habitat preference of North American pachycephalosaurids (Dinosauria, Ornithischia)'. *Lethaia*, 47(4), pp. 567-578.

Rich, T.H., Vickers-Rich, P. and Gangloff, R.A., 2002. 'Polar dinosaurs'. *Science*, 295(5557), pp. 979-980.

Sullivan, C., Wang, Y., Hone, D.W., Wang, Y., Xu, X. and Zhang, F., 2014. 'The vertebrates of the Jurassic Daohugou Biota of northeastern China'. *Journal of Vertebrate Paleontology*, 34(2), pp. 243-280.

Varricchio, D.J., Martin, A.J. and Katsura, Y., 2007. 'First trace and body fossil evidence of a burrowing, denning dinosaur'. *Proceedings of the Royal Society of London B: Biological Sciences*, 274(1616), pp. 1361-1368.

Chapter 7: Anatomy

Bell, P.R., Fanti, F., Currie, P.J. and Arbour, V.M., 2014. 'A mummified duck-billed dinosaur with a soft-tissue cock's comb'. *Current Biology*, 24(1), pp. 70-75.

Bonnan, M.F., Sandrik, J.L., Nishiwaki, T., Wilhite, D., Elsey, R.M. and Vittore, C., 2010. 'Calcified cartilage shape in archosaur long bones reflects overlying joint shape in stress-bearing elements: implications for nonavian dinosaur locomotion'. *The Anatomical Record*, 293(12), pp. 2044-2055.

Dal Sasso, C. and Signore, M., 1998. 'Exceptional soft-tissue preservation in a theropod dinosaur from Italy'. *Nature*, 392(6674), p. 383.

Hone, D.W., Persons, W.S. and Le Comber, S.C., 2021. 'New data on tail lengths and variation along the caudal series in the non-avialan dinosaurs'. *PeerJ*, 9, p. e10721.

Lee, Y.N., Barsbold, R., Currie, P.J., Kobayashi, Y., Lee, H.J., Godefroit, P., Escuillié, F. and Chinzorig, T., 2014. 'Resolving the long-standing enigmas of a giant ornithomimosaur *Deinocheirus mirificus*'. *Nature*, 515(7526), p. 257.

Morhardt, A.C., 2009. *Dinosaur smiles: Do the texture and morphology of the premaxilla, maxilla, and dentary bones of sauropsids provide osteological correlates for inferring extra-oral structures reliably in dinosaurs?*. Western Illinois University.

Chapter 8: Mechanics and Movement

Bates, K.T., Falkingham, P.L., Macaulay, S., Brassey, C. and Maidment, S.C., 2015. 'Downsizing a giant: re-evaluating *Dreadnoughtus* body mass'. *Biology Letters*, 11(6), p. 20150215.

Hutchinson, J.R., Ng-Thow-Hing, V. and Anderson, F.C., 2007. 'A 3D interactive method for estimating body segmental parameters in animals: application to the turning and running performance of *Tyrannosaurus rex*'. *Journal of Theoretical Biology*, 246(4), pp. 660-680.

Mallison, H., 2010. 'The digital *Plateosaurus* II: an assessment of the range of motion of the limbs and vertebral column and of previous reconstructions using a digital skeletal mount'. *Acta Palaeontologica Polonica*, 55(3), pp. 433-458.

Persons IV, W.S. and Currie, P.J., 2011. 'Dinosaur speed demon: the caudal musculature of *Carnotaurus sastrei* and implications for the evolution of South American abelisaurids'. *PloS one*, 6(10), p. e25763.

Taylor, M.P., Wedel, M.J. and Naish, D., 2009. 'Head and neck posture in sauropod dinosaurs inferred from extant animals'. *Acta Palaeontologica Polonica*, 54(2), pp. 213-220.

Chapter 9: Physiology

Eagle, R.A., Tütken, T., Martin, T.S., Tripati, A.K., Fricke, H.C., Connely, M., Cifelli, R.L. and Eiler, J.M., 2011. 'Dinosaur body temperatures determined from isotopic (13C-18O) ordering in fossil biominerals'. *Science*, 333(6041), pp. 443-445.

Erickson, G.M., Makovicky, P.J., Currie, P.J., Norell, M.A., Yerby, S.A. and Brochu, C.A., 2004. 'Gigantism and comparative life-history parameters of tyrannosaurid dinosaurs'. *Nature*, 430(7001), p. 772.

Hummel, J., Gee, C.T., Südekum, K.H., Sander, P.M., Nogge, G. and Clauss, M., 2008. 'In vitro digestibility of fern and gymnosperm foliage: implications for sauropod feeding ecology and diet selection'. *Proceedings of the Royal Society of London B: Biological Sciences*, 275(1638), pp. 1015-1021.

Köhler, M., Marín-Moratalla, N., Jordana, X. and Aanes, R., 2012. 'Seasonal bone growth and physiology in endotherms shed light on dinosaur physiology'. *Nature*, 487(7407), p. 358.

Leuzinger, L., Bernasconi, S.M., Vennemann, T., Luz, Z., Vonlanthen, P., Ulianov, A., Baumgartner-Mora, C., Hechenleitner, E.M., Fiorelli, L.E. and Alasino, P.H., 2021. 'Life and reproduction of titanosaurians: isotopic hallmark of mid-palaeolatitude eggshells and its significance for body temperature, diet, and nesting'. *Chemical Geology*, p. 120452.

Chapter 10: Coverings

Arbour, V.M., Lech-Hernes, N.L., Guldberg, T.E., Hurum, J.H. and Currie, P.J., 2013. 'An ankylosaurid dinosaur from Mongolia with in situ armour and keratinous scale impressions'. *Acta Palaeontologica Polonica*, 58(1), pp. 55-64.

Bell, P.R., Campione, N.E., Persons, W.S., Currie, P.J., Larson, P.L., Tanke, D.H. and Bakker, R.T., 2017. 'Tyrannosauroid integument reveals conflicting patterns of gigantism and feather evolution'. *Biology letters*, 13(6), p. 20170092.

Godefroit, P., Sinitsa, S.M., Dhouailly, D., Bolotsky, Y.L., Sizov, A.V., McNamara, M.E., Benton, M.J. and Spagna, P., 2014. 'A Jurassic ornithischian dinosaur from Siberia with both feathers and scales'. *Science*, 345(6195), pp. 451-455.

Lovelace, D.M., Hartman, S.A., Mathewson, P.D., Linzmeier, B.J. and Porter, W.P., 2020. 'Modeling Dragons: using linked mechanistic physiological and microclimate models to explore environmental, physiological, and morphological constraints on the early evolution of dinosaurs'. *PLoS one*, 15(5), p. e0223872.

Platt, B.F. and Hasiotis, S.T., 2006. 'Newly discovered sauropod dinosaur tracks with skin and foot-pad impressions from the Upper Jurassic Morrison Formation', Bighorn Basin, Wyoming, USA. *Palaios*, 21(3), pp. 249-261.

Chapter 11: Appearance

Brown, C.M., Henderson, D.M., Vinther, J., Fletcher, I., Sistiaga, A., Herrera, J. and Summons, R.E., 2017. 'An exceptionally preserved three-dimensional armored dinosaur reveals insights into coloration and Cretaceous predator-prey dynamics. *Current Biology*, 27(16), pp. 2514-2521.

Li, Q., Gao, K.Q., Vinther, J., Shawkey, M.D., Clarke, J.A., D'alba, L., Meng, Q., Briggs, D.E. and Prum, R.O., 2010. 'Plumage color patterns of an extinct dinosaur'. *Science*, 327(5971), pp. 1369-1372.

Hu, D., Clarke, J.A., Eliason, C.M., Qiu, R., Li, Q., Shawkey, M.D., Zhao, C., D'Alba, L., Jiang, J. and Xu, X., 2018. 'A bony-crested Jurassic dinosaur with evidence of iridescent plumage highlights complexity in early paravian evolution'. *Nature communications*, 9(1), p. 217.

Xu, X., Zheng, X. and You, H., 2010. 'Exceptional dinosaur fossils show ontogenetic development of early feathers'. *Nature*, 464(7293), p. 1338.

Chapter 12: Reproduction

Erickson, G.M., Makovicky, P.J., Inouye, B.D., Zhou, C.F. and Gao, K.Q., 2009. 'A life table for *Psittacosaurus lujiatunensis*: initial insights into orni-thischian dinosaur population biology'. *The Anatomical Record*, 292(9), pp. 1514-1521.

Erickson, G.M., Zelenitsky, D.K., Kay, D.I. and Norell, M.A., 2017. 'Dinosaur incubation periods directly determined from growth-line counts in embry-onic teeth show reptilian-grade development'. *Proceedings of the National Academy of Sciences*, 114(3), pp. 540-545.

Hone, D.W., Farke, A.A., Watabe, M., Shigeru, S. and Tsogtbaatar, K., 2014. 'A new mass mortality of juvenile *Protoceratops* and size-segregated aggregation behaviour in juvenile non-avian dinosaurs'. *PloS one*, 9(11), p. e113306.

Isles, T.E., 2009. 'The socio-sexual behaviour of extant archosaurs: implica-tions for understanding dinosaur behaviour'. *Historical Biology*, 21(3-4), pp. 139-214.

Meng, Q., Liu, J., Varricchio, D.J., Huang, T. and Gao, C., 2004. 'Palaeontology: parental care in an ornithischian dinosaur'. *Nature*, 431(7005), p. 145.

Norell, M.A., Clark, J.M., Chiappe, L.M. and Dashzeveg, D., 1995. 'A nesting dinosaur'. *Nature*, 378(6559), p. 774.

Chapter 13: Behaviour

Hone, D.W., Naish, D. and Cuthill, I.C., 2012. 'Does mutual sexual selection explain the evolution of head crests in pterosaurs and dinosaurs?'. *Lethaia*, 45(2), pp. 139-156.

Tanke, D.H. and Currie, P.J., 1998. 'Head-biting behavior in theropod dino-saurs: paleopathological evidence'. *Gaia*, 15, pp. 167-184.

Terrill, D.F., Henderson, C.M. and Anderson, J.S., 2020. 'New application of strontium isotopes reveals evidence of limited migratory behaviour in Late Cretaceous hadrosaurs'. *Biology Letters*, 16(3), p. 20190930.

Witmer, L.M., Ridgely, R.C., Dufeau, D.L. and Semones, M.C., 2008. 'Using CT to peer into the past: 3D visualization of the brain and ear regions of birds, crocodiles, and nonavian dinosaurs'. In *Anatomical imaging* (pp. 67-87). Springer, Tokyo.

Witmer, L.M. and Ridgely, R.C., 2009. 'New insights into the brain, braincase, and ear region of tyrannosaurs (Dinosauria, Theropoda), with implications for sensory organization and behavior'. *The Anatomical Record*, 292(9), pp. 1266-1296.

Chapter 14: Ecology

Barrett, P.M. and Rayfield, E.J., 2006. 'Ecological and evolutionary implications of dinosaur feeding behaviour'. *Trends in Ecology & Evolution*, 21(4), pp. 217-224.

Barrett, P.M. and Willis, K.J., 2001. 'Did dinosaurs invent flowers? Dinosaur–angiosperm coevolution revisited'. *Biological Reviews*, 76(3), pp. 411-447.

Mallon, J.C., Evans, D.C., Ryan, M.J. and Anderson, J.S., 2013. 'Feeding height stratification among the herbivorous dinosaurs from the Dinosaur Park Formation (upper Campanian) of Alberta, Canada'. *BMC Ecology*, 13(1), p. 14.

Schmitz, L. and Motani, R., 2011. 'Nocturnality in dinosaurs inferred from scleral ring and orbit morphology'. *Science*, 332(6030), pp. 705-708.

Witzmann, F., Claeson, K.M., Hampe, O., Wieder, F., Hilger, A., Manke, I., Niederhagen, M., Rothschild, B.M. and Asbach, P., 2011. 'Paget disease of bone in a Jurassic dinosaur'. *Current Biology*, 21(17), pp. R647-R648.

Chapter 15: Dinosaur Descendants

Bhullar, B.A.S., Marugán-Lobón, J., Racimo, F., Bever, G.S., Rowe, T.B., Norell, M.A. and Abzhanov, A., 2012. 'Birds have paedomorphic dinosaur skulls'. *Nature*, 487(7406), p. 223.

Dial, K.P., Jackson, B.E. and Segre, P., 2008. 'A fundamental avian wing-stroke provides a new perspective on the evolution of flight'. *Nature*, 451(7181), p. 985.

Xu, X., Zhou, Z., Wang, X., Kuang, X., Zhang, F. and Du, X., 2003. 'Four-winged dinosaurs from China'. *Nature*, 421(6921), p. 335.

Xu, X., Zhou, Z., Dudley, R., Mackem, S., Chuong, C.M., Erickson, G.M. and Varricchio, D.J., 2014. 'An integrative approach to understanding bird origins'. *Science*, 346(6215), p. 1253293.

Zelenitsky, D.K., Therrien, F., Erickson, G.M., DeBuhr, C.L., Kobayashi, Y., Eberth, D.A. and Hadfield, F., 2012. 'Feathered non-avian dinosaurs from North America provide insight into wing origins'. *Science*, 338(6106), pp. 510-514.

Chapter 16: Research and Communication

Milner, A.R., Harris, J.D., Lockley, M.G., Kirkland, J.I. and Matthews, N.A., 2009. 'Bird-like anatomy, posture, and behavior revealed by an Early Jurassic theropod dinosaur resting trace.' *PloS one*, 4(3), p. e4591.

Schweitzer, M.H., Marshall, M., Carron, K., Bohle, D.S., Busse, S.C., Arnold, E.V., Barnard, D., Horner, J.R. and Starkey, J.R., 1997. 'Heme compounds in dinosaur trabecular bone'. *Proceedings of the National Academy of Sciences*, 94(12), pp. 6291-6296.

Sullivan, C., Hone, D. and Xu, X., 2012. 'The search for dinosaurs in Asia'. *The Complete Dinosaur*, pp. 61-73.

Wang, C.C., Song, Y.F., Song, S.R., Ji, Q., Chiang, C.C., Meng, Q., Li, H., Hsiao, K., Lu, Y.C., Shew, B.Y. and Huang, T., 2015. 'Evolution and function of dinosaur teeth at ultramicrostructural level revealed using synchrotron transmission X-ray microscopy.' *Scientific reports*, 5, p. 15202.

Xing, L., McKellar, R.C., Xu, X., Li, G., Bai, M., Persons IV, W.S., Miyashita, T., Benton, M.J., Zhang, J., Wolfe, A.P. and Yi, Q., 2016. 'A feathered dinosaur tail with primitive plumage trapped in mid-Cretaceous amber'. *Current Biology*, 26(24), pp. 3352-3360.

If you enjoyed this book . . .

. . . and hopefully you did, then this is my opportunity to plug some of my other ongoing ways of reaching out and communicating about science. I would ask that if this book (or my other work) has at all inspired you to go on and become more interested in dinosaurs to please take the literal two minutes to fill in this form: (https://www. davehone.co.uk/outreach/impact-survey/). Keeping track of what impact my work has on the general public helps me to keep doing it.

If you do want to see more of my work, then there's lots of links to things that I have done through my website: https://www.davehone. co.uk. My previous book, *The Tyrannosaur Chronicles* (Bloomsbury, 2016) on the evolution and biology of the tyrannosaurs, is still in print. If you want to hear my voice, listen to a podcast on dinosaurs, *Terrible Lizards*, run with my friend Iszi Lawrence (https://terriblelizards. libsyn.com) and if you want to see me, on YouTube I have a playlist of all my talks and interviews (https://tinyurl.com/2ux6ct34).

You can follow me on Twitter as @Dave_Hone and my old blog is just about still limping on (https://archosaurmusings.wordpress.com/); while the updates might be few and far between, there are now some fifteen years of posts on there, so there is plenty to read.

Image Permissions

Text

P7. Illustration by Scott Hartman.

P15. Image by and courtesy of David Evans.

P26. Illustration by Scott Hartman.

P32. Illustration by Scott Hartman.

P58. Illustration by Scott Hartman.

P76. Illustration by Scott Hartman.

P94. I'm the photographer / owner.

P100. Image by Jordan Mallon. Courtesy of the Canadian Museum of Nature.

P121. Image by and courtesy of Donald Henderson.

P127. Image by and courtesy of Jordan Mallon.

P132. Image by and courtesy of David Evans.

P147. Image by and courtesy of Pascal Godefroit.

P161. Image by John Scanella. Courtesy of the Museum of the Rockies.

P170. I'm the photographer / owner.

P179. I'm the photographer / owner.

P200. Image from Wolff , E.D.S., Salisbury, S.W., Horner, J.R. and Varricchio, D.J. (2009). 'Common Avian Infection Plagued the Tyrant Dinosaurs'. *PLoS ONE*, 4(9), pp. e7288.)

P208. Image by and courtesy of Julia Clarke.

P220. Image by and courtesy of Evan Johnson-Ransom and Eric Snively.

P228 Image by Matt Zeher. Image courtesy of Lindsay Zanno and the North Carolina Museum of Natural Sciences.

Photo Insert

1. Image by Sven Tränkner and courtesy of Philipe Havlik and the Senckenberg Museum.
2. Image by and courtesy of Xing Lida.
3. Image by and courtesy of Xu Xing.
4. Image courtesy of Caleb Brown and the Royal Tyrrell Museum of Palaeontology.
5. Image by and courtesy of Maria McNamara.
6. Image by and courtesy of David Evans.
7. Image by Thierry Hubin and courtesy of Andrea Cau and the Royal Belgian Institute of Natural Science.
8. Image by Mick Ellison and courtesy of the American Museum of Natural History.
9. Image by and courtesy of Martin Kundrát.
10. Image by and courtesy of Lawrence Witmer.
11. Image by and courtesy of Heinrich Mallison.
12. Image by and courtesy of Peter Falkingham.
13. Image by and courtesy of ReBecca Hunt-Foster.
14. Image by and courtesy of Alejandro Otero.

Index

Page numbers in **bold** refer to illustrations.

A
abelisaurs 59, 61, 70, 186
activity times 95
Africa 24, 66, 89–90, 91, 194–5; *see also specific country*
'Age of Reptiles' name 12
agility 117–18
Ajkaceratops 83
Alamosaurus 89
alligators 34, 93, 113, 127, **127**, 143–4, 177
allosaurs 59, 107, 145
Allosaurus 8, 30, 182, 185
Alvarez, Luis and Walter 14
alvarezsaurs 61, 70, 79, 121, 159, 197
Amargasaurs 64
amber 217
amniotes 27
Amphicoelias 51, **52**
anatomy 97, 107–8
  data from bones 98–9
  digestive systems 105–6
  fat deposits 105
  function of features 106–7
  furculae 101
  importance 98–9

joints 110–11
mechanics and movement *see* biomechanics; movement
  missing data problems 102
  muscle composition 104–5
  retrodeformation 99–100
  skulls 99, **100**
  small bones, missing 100
  tails *see* tails
Anchiornis 153–6, 160
ankylosaurs 46, 65, 66, **121**, 156, 166, 182–3, 190
Ankylosaurus 32, 77
apatosaurs 64
Apatosaurus 64
appearance
  colouration and patterning *see* colouration and patterning
  and communication 181–2
  crests 149, 160–2
  Edmontosaurus 160, **161**
  eye shape and colour 158–9
  eyelashes 159
  facial discs 159
  feathers *see* feathered dinosaurs; feathers

mouth bristles 159
wattles and combs 160–2
Archaeopteryx 30, 34, 43, 53, 101, 195, 205, 215, 216
archosaur group 28–9, 30, 216
Arctic and Antarctic regions 25, 93, 105, 207
Argentinosaurus 77
armour 136, 137–8
arms 107, 109, 186
Asia 82, 83; *see also specific country*
asteroid extinction theory (K-Pg) 14–17, **15**

B
Balaur 205
beaks 62, 141–2, 206
behaviours
  carnivores 183–6
  communication 181–3
  competition 181–2
  daily rhythms 186–7
  feeding ecology 180–1
  fighting 167–8, 182–3
  future discoveries 227
  herbivores 182–3
  intelligence 176–7
  social 177–81, **179**
Belgium 6–8
biological species concept 69
biomechanics 109, 122
  agility 117–18
  climbing 119–20
  digging 120–1
  feeding 190
  flight, powered 118–19
  future discoveries 227

jumping 119
limits of large size 112–14
mating 117
necks 115–16
posture 116–17
swimming and floating 120, **121**
tails 115
walking and running 117–18
weight 110–12
bipedalism 23, 30, 31, 62, 65, 81, 101, 116–17
birds
  ancient representatives 206–9
  appearance 159–60
  diversification 81
  evolutionary pathway 30, 31, 33–4, 79–80, 205–6, 215–16
  feathers 127, 128, 142–3
  furculae 101
  intelligence 176
  K–Pg event, impact of 19–20, 209–10
  living dinosaurs x*n*, 203
  Mesozoic 62
  muscle composition 104
  number of species 55
  paedomorphosis 79–80
  powered flight 118–19, 203–5
  and pterosaurs 207–9
  reproduction 163–4, 168, 169–70, 171
  scales 143
  thermophysiology 127, 128
  Vegavis 206–7, **208**
Boreaopelta 156
'bottom-up' effect 197–8
brachiosaurs 33, 64

Brachiosaurus 30, 31, 33, 64
brain size and structure 176–7,
    187
Brazil 8, 10, 24–5, 27, 142
breathing, circular 217
Brontosaurus 31, 64
Buckland, William 13

C
camouflage 94–5, 150, 152–3, 155,
    157
CAMP (Central Atlantic Magmatic
    Province) volcanic events
    29–30
Canada **15**, 44, 89, 142, 160, **161**,
    216
Carcharodontosaurus 195
carnivores *see* theropods
Centrosaurus **100**
ceratopsians 67, 77, 83, 99, **100**, 104,
    120, 157, 166, 167; *see also*
    specific species
ceratosaurs 59, 141–2
Ceratosaurus 59
chimeras 50
China
    Anchiornis 153
    feathered dinosaurs 10, 44, 119,
        142, 203, 216
    horned dinosaurs 83
    Lystrosaurus 25, 63
    nests and eggs **170**
    ornithischians 83
    oviraptorosaurs **170**
    pterosaurs 90
    research investment 214
    Shantungosaurus 49

Sinosauropteryx 151–2
sites 90, 93, 103, 194, 216
stegosaurs 80
trackways **179**
tree trunk fossils 93, **94**
clades 30–2, **32**
claws 107, 138–9, 186, 190–1
climate 41, 90–1, 95–6, 105, 124,
    126, 127, 128
climbing 119–20
coelophysids 59
Coelophysis 59
colouration and patterning 9–10,
    162
    age differences 155
    Anchiornis 153–6
    camouflage 95, 152–3, 157
    change, ability to 156
    countershading 152–3
    environmental context 155
    eyes 158
    function of colour 156–61
    future discoveries 155–6
    gender differences 154–5
    melanosomes 150–2
    research difficulties 152
    scales 156
    Sinosauropteryx 151–2
    stereotypes 149–50
    variation within species 154
    warning and mimicry 157–8
combs 160–2
communication, animal 135, 181–3
communication of research
    218–22
competition 74, 78–9, 84, 181–2,
    192–3, 197

compsognathids 60, 151–2, 195
Compsognathus 60
computer modelling 44, 46, 109,
    111, **121**, 218–20, **220**
cooling strategies 128–9
Cope, E.D. 51
courtship and mating 117, 166–9
coverings 147–8
    armour 137–8
    beaks 141–2
    claws 138–9
    feathers *see* feathered dinosaurs;
      feathers
    filaments 145–7
    keratin 137–8
    lips 139–41
    new discoveries 135
    scales 135, 136–9, 143–4, **147**
    skin 135, 136–9
    spikes 137–8
    unguals 138
cranks 222*n*
crests 149, 160–2
Cretaceous Period 12–13
crocodiles 3, 4, 34, 89, 125, 126, 140,
    156, 171, 174, 195, 217
crocodilians 28–9, 34, 89–90, 126–7,
    140, 149, 161, 177, 195
CT scans 213, 218
Cuvier, Baron Georges 13

D
daily rhythms 186–7
Darwin, Charles 2
decay 41–2, 71
Deccan Trap, India 18
Deinocheirus 77

deserts 42, 105
dicraeosaurs 64
diet 3–4, 81–2, 105–6, 130–1, 190–3,
    194–5, 196–7
digestive systems 105–6, 130
digging 120–1
Dilong **220**
'Dinosauria' name 2, 6
dinosauromorphs 24–7, **26**, 28, 74
diplodocids 63–4
Diplodocus 8, 13, 30, 63, 138, 190
diseases 131–2, 199–201, **200**
diversity **58**
    compared to living animal groups
      56–7, 67
    discoveries 67–8
    evolutionary patterns 80–2
    islands 70–1
    missing species 70–2, 103
    number of species 55, 68–9
    ornithischians 65–7
    rainforests 71
    sauropods 62–5
    species, recognition of 69–70
    theropods 57, 59–62
    undiscovered species 71–2
domination 27–30
Doyle, Sir Arthur Conan 13*n*
dromaeosaurs 33–4, 62, 79, 105,
    138–9, 159–60, 186, 204, 205
drones 222
droughts 77, 96, 194

E
ears 187
ecology 189, 201
    diet 190–3

ecosystems 193–6
future discoveries 227–8
interactions 197–9
niche partitioning 192–3
parasites and diseases 131–2,
    199–201, **200**
population density 192
population regulators 198
specialist/generalist feeding
    196–7
'top-down' and 'bottom-up'
    effects 197–8
ecosystem control 197–8
ecosystems 92–4, **94**, 192–6
Edmontosaurus 160, **161**
eggs 164, 169–72, **170**
encephalisation quotient (EQ)
    176–7
England *see* United Kingdom
environment *see* climate; ecosystems;
    habitats
Euoplocephalus **121**
Europe 83, 226; *see also specific
    country*
evolution, theory of 2
evolutionary-developmental
    (evo-devo) biology 215,
    217–18
evolutionary patterns
    co-evolution with plants 82
    diversity increasing 80–2
    environmental impacts 73
    locations 82–3
    size, changes in 74–80, **76**
    species duration 84–5
    species extinctions 84
    speed 74

evolutionary relationships 30, 33–8,
    145
extinction, Permian-era 27
extinction, survival of 19–21, 75, 78,
    209–10
extinction theories, mass 11–12
    asteroid theory 14–17, **15**
    infection 199–200
    Victorian Era 12, 13–14
    volcanic destruction theory
        18–19
extinction, Triassic-period 28,
    29–30
eyelashes 159
eyes 158–9, 186

F
face scales 139
facial discs 159
family trees 33–6
fat deposits 105
feathered dinosaurs 30, 43, 53, 60,
    102, 107, 128, 142–3, 144–5,
    147–8, 215–16
feathers 128, 142–7, 159–60,
    215–16
fighting 167–8, 182–3
filaments 145–8, **147**, 216–17
flight, powered 118–19, 203–5
floods 96
footprints 45–8, 136
fossilisation 40, 41–2
frills 99, 166
furculae 101
future discoveries 26, 71–2, 85, 117,
    180, 215–18, 226–8

G
Galloanseriformes 207
gastroliths 191
geology – palaeontology links 2, 12
Germany 43, 51, 53, 59, 63, 142,
    195–6, 216
giantism 77*n*
gigantohomeothermy 125, 129–30
Gigantoraptor 77
Gigantosaurus 77
Giraffatitan 64, 110
gliders 34, 62, 79, 93, 101–2, 120,
    156, 159–60, 204
growth rate 125–6, **127**

H
habitats 87
  activity times 95
  climatic and seasonal influences
    90–1, 95–6
  colour 95
  and dinosaur size 88–92
  dinosaur ubiquity across 89–90
  ecosystem changes 90–1
  ecosystems 92–4, **94**
  light 94–5
hadrosaurs 32, 66–7, 77, 91, 116, **132**,
    136, 137, 160, 170, 172, 196
hearing 187
Hennig, Willi 35
herbivores 30, 81–2; *see also*
    ornithischians; sauropods
Hesperornithes 210
heterodontosaurs 65
heterotherms 123–4, 127–8
hibernation 93, 94, 124, 127
hips 24, 31

homeotherms 123, 124, 125, 130
horns 74, 137, 167
Hungary 83
hunting/scavenging 183–6, 192,
    198
Huxley, Thomas Henry 36
Hylaeosaurus 6

I
ichnology 47
Iguanodon 3–6, 6–8, **7**, 12–13, 66
iguanodontids 66, 137, 138, 191
India 8, 18
Indonesia 18
injuries and infections 131–2, **132**,
    199–201, **200**
integument 135
intelligence 176–7
iridium 14, 15, **15**
islands 70–2, 195–6, 205
isotope analysis 91–2
Italy 44, 105–6

J
Jinguofortis 206
jumping 119
Jurassic Park series 59, 132, 149, 199
Jurassic Period 12

K
K-Pg extinction theory 14–17, **15**
keratin 137–8, 141, 143, 156
Keynes, John Maynard xii
Kulindadromeus 146–7, **147**

L
Lagerstaetten beds 43–4, 100*n*, 159

light 94–5
Limusaurus 141–2
lips 139–42
lizards 31, 126, 139, 156, 163–4, 217
Lufengosaurus 63
Lystrosaurus 25

M
Mallon, Jordan 164*n*
mamenchisaurs 63
Mamenchisaurus 103, 116
Mantell, Gideon 3–5, 13
Mantellisaurus 8*n*
mass mortalities 49–50
mass mortality sites 178–9, 185, 194, 199
mass/weight 45, 75, 77, 110–12, 113*n*, 114, 115
mating 117, 166–9
medullary bones 164
megalosaurs 59–60
Megalosaurus 3, 6, 13
melanosomes 150–2, 156
Mesozoic Era 12–13
Mexico 15–16, 89
Microraptor 102
migration 88–9, 90–2, 180
mimicry 157–8
molecular palaeontology 226–7
Mongolia 8, 61, 144–5, 180
mouth bristles 159
movement 77–8, 101, 109, 117–22; *see also* biomechanics
muscle composition 104–5
museums 50–4, 219, 220
Myanmar 217

N
naming rules 61*n*
necks 31, 115–16, 129
nests and eggs 164, 169–72, **170**
niche partitioning 192–3
Niger 89–90, 104
Nigersaurus 190
North Africa 194–5
North America 8, 15, 64, 82, 83, 89, 127, 144–5, 226; *see also* Canada; United States
North Korea 142
number of species 55, 68–9
Nyasasaurus 24, 25, 27

O
On the Origin of Species (Darwin) 2
origins 223
  difficulties determining 23
  dinosauromorphs 24–7, **26**
  evolution and domination 27–30
  evolutionary relationships 30
  first dinosaurs 23–7, **26**
  Nyasasaurus 24, 25, 27
ornithischian–saurischian split 36–8
ornithischians **32**; *see also specific group*
  beaks 141
  camouflage 156
  claws 138
  climbing 120
  digging 121
  diversity 65–7
  filaments 145–7, **147**, 148, 216–17
  finds 49, 103
  groups 65–7
  origins 83

ornithischian–saurischian split
36–8
overview 32
pubis 31
scales **147**, 148
size 49, 75
tails 115
teeth 142
ornithomimosaurs 60, 77, 104
ornithopods 66
Ornithoscleida 36–8
Oryctodromeus 121
Ouranosaurus 66, 105
Oviraptor 142, 191
oviraptorosaurs 61, 70, 77, 79, 142,
159, **170**, 191
Owen, Richard 6, 8

P
pachycephalosaurs 67
paedomorphosis 79–80
palaeognaths 207
palaeontology, early 1–9
palynology 92–3
Pangea 25
Parasaurolophus **132**
parasites and diseases 131–2,
199–201, **200**
parental care 172–4
patterning *see* colouration and
patterning
Pelicanimimus 60
Permian-era mass extinction 27
phalluses 169
photogrammetry 218–19
physiology 133
digestion 130

growth rate 125–6, **127**
injuries and infections 131–2,
132
temperature *see*
thermophysiology
venom 132–3
Pisanosaurus 83
plant ecosystems 92–4, **94**
plant evolution 82, 87
Plateosaurus 63
pneumatic bones 111
pollen, fossil 92–3
posture 4, 28, 115, 116–17
preservation of remains
burial 42
climatic events 95–6
data limitations 39–40
environmental conditions 41–2
footprints 45–8
Lagerstaetten-type deposits 43–4
movement of remains 48–50
museums 50–4
requirements 41
soft tissues 43–4
taphonomic history, tracing
48–50
prosauropods 31, 62, 81, 101, 116
Protoceratops 180
Psittacosaurus 146, 156
Pteranodon 209
pterosaurs 13, 28–9, 90, 118, 145*n*,
195, 207–9
public engagement 220–2

Q
quadrupedalism 23, 31, 62, 65, 81,
102, 116–17

quartz, shocked 15, 16

R
rainforests 41, 42, 71
rebbachisaurids 64
record keeping 53
reproduction 163, 174
    and communication 182
    courtship and mating 117, 166–9
    egg-laying 164
    evolutionary and ecological
        impacts 173
    growth and maturity 174
    males and females 163–6
    nests and eggs 164, 169–72, **170**
    parental care 172–4
    phalluses 169, 169*n*
    sexual dimorphism 164–6
research; *see also* research technology
    communication of results
        218–22
    evolutionary-developmental
        (evo-devo) biology 215,
        217–18
    finding and accessing relevant
        papers 211–12
    funding 98, 214, 220, 222
    future discoveries 215–18
    new fields 215
    non-academic researchers 222
    public engagement 220–2
    researcher numbers 68, 214
research technology
    3D imaging and printing
        218–19, **220**
    communication 212, 218, 220–1
    computer modelling 44, 46, 109,

        111, **121**, 122, 218–20, **220**
    drones 222
    genetics 35
    molecular palaeontology 226–7
    photogrammetry 218–19
    scans of fossils 40, 99–100, 152,
        187, **208**, 213–14, 218, **220**
Romania 70, 205
Russia 25, 146–7, 216–17

S
saltasaurs 65
saurischians 31; *see also*
    sauropodomorphs; sauropods;
    theropods
Sauropelta **121**
sauropodomorphs 30, 31, **32**, 36, 51,
    62–3, 75, 101, 138, 148; *see also*
    prosauropods; *specific group*;
    specific species
sauropods 36–7; *see also specific*
    species
    claws 120, 138
    diversity 63–5
    epoch 77
    evolutionary relationships 33, 81
    feeding ecology 190, 192–3
    finds 90, 99, 103
    footprints 46
    groups 63–5
    intelligence 176, 177
    migration 91
    muscle composition 104
    necks 115–16
    new discoveries 113*n*
    posture 116–17
    size 31, 112, 113

thermophysiology 129–30

scales 135, 136–9, 143–4, 146–8, **147**, 156, **161**

scans of fossils 40, 99–100, 152, 187, **208**, 213–14, 218, **220**

scansoriopterygids 61–2, 70, 204

scavenging 183–6

Scipionyx 44, 105–6

senses 158–9, 186–7

sexual dimorphism 164–6

Shantungosaurus 49, 77

Shunosaurus 63, 103

Sinornithosaurus 132

Sinosauropteryx 151–2

size 3, 30, 31, 32, 74–80, **76**, 81, 88–92, 112–14; *see also* mass/ weight

skin 135, 136–9, 144–5

smell, sense of 187

snakes 124, 126, 156, 163–4

social behaviours 177–81, **179**

soft tissues 40, 43–4, 47–8, 104–6, 217

South Africa 25, 83

South America 8, 24–5, 64, 65, 70, 83; *see also specific country*

specialist/generalist feeding 196–9

species duration/extinction 84–5

species, missing 43, 67–70, 71, 103

species, number of 9, 55–6, 68–9

species, recognition of 69, 103

speed 4, 117–18

spikes 137–8

Spinophorosaurus 104

spinosaurs 59–60, 81, 102

Spinosaurus 30, 51, 59, 77, 102, 105, 194, 195

stegosaurs 65–6, 80, 84

Stegosaurus 8, 32, 65–6, 193

swimming and floating 120, **121**

T

tails 31, 101–2, 103–4, 115, 159–60, **161**, 182–3

Tanzania 8, 24, 64

taphonomy 40–50

technology *see* research technology

teeth 3–4, 91–2, 141, 142, 190, 191

temperature, body *see* thermophysiology

Thecodontosaurus 51

therizinosaurs 60–1, 139

thermophysiology
age-related 129–30
cooling strategies 128–9
and dietary needs 130–1
gigantohomeothermy 125, 129–30
hibernation and torpor 124
high body temperatures 126–8
homeotherms and heterotherms 123–4
surface-area-to-volume ratios 124–5

theropods **32**; *see also specific group; specific species*
anatomy 31, 101, 107, 138, 141–2
ancestors of birds 31, 33–4, 79–80, 205–6
climbing 119
courtship and mating 167, 168

diet 194–5
diversity 57, 59–62
feathers 142–3, 144–5, 147, 159,
   160
groups 59–62
herbivory, switch to 81–2
hunting/scavenging 183–6,
   198
intelligence 176
on islands 195
Late Cretaceous period 77
nests and eggs 170
Ornithoscleida proposal 36–8
overview 30
size 75, 77, 79, 112–13
social behaviours **179**
thermophysiology 127, 128
3D imaging and printing 99,
   218–19, **220**
thyreophoreans 65–6
titanosaurs 64, 65, 70, 77, 89, 171
'top-down' effect 197–8
torpor 124
trace fossils 47
trackways 45–7, 115, 178–80, **179**
Transylvania 70
Triassic Period 12
Triassic-period extinction 28,
   29–30
Triceratops 8, 10, 13, 32, 67, 77, 146,
   167
Trichomonas 199
troodontids 33–4, 62, 79, 153–5,
   160, 176, 186, 187, 190–1, 204,
   205
tyrannosaurs
   3D scans **220**

anatomy 107, 141
feathers and/or scales 142, 144–5,
   147
finds **228**
hunting 184, 186, 191–2
injuries and infections 167,
   182–3, 199, **200**
lips 141
muscle composition 104–5
overview 60
size 77–8
thermophysiology 128
tracing ancestry of 82
Tyrannosaurus
   epoch 77
   feathers and/or scales 144–5
   finds 47–8, 99
   geographic range 89
   injuries and infections **200**
   intelligence 176*n*
   sexual dimorphism 164*n*
   size 111, 113, 114
   smell, sense of 187*n*
   speed and agility 117–18
   T. rex ix–x, 111, **200**

U
unguals 138
United Kingdom 1, 3–6, 25, 27, 51,
   82, 83, 220–1, 223
United States 25, 27, 64, 89, 113,
   168, 193–4, 226

V
Vegavis 206–7, **208**
Velociraptor 30, 105, 142
Venezuela 25

venom 132–3

vision 158–9, 186

volcanic extinction events 18–19, 28, 29–30

W

warning colouration/patterning 157–8

wattles 160–2

weight/mass 45, 75, 77, 110–12, 113n, 114, 115

X

X-ray scanning 152, 187, 213

X-ray videos 46

Y

Yi 62